Building a Ferromagnetic Generator

By

John Boyle

I'd like to especially thank the great artist who created my book cover:

Imran Shaika (Retina99)

www.retina99.com

This book is dedicated to the memory of my wife:
Deborah Kathleen Woolley.

<u>All rights reserved.</u>

<u>Please Note: This book describes some dangerous practices which should not be copied or attempted by anyone who does not understand the underlying physics and the limits of the materials used.</u>

The author assumes no liability for any damages or injuries resulting from the use of the information contained herein.

First Edition.

Copyright © 2017 by John Boyle

Introduction

Building a Ferromagnetic Generator.

    This project began back in nineteen eighty-eight when I became housebound and needed a way to keep my mind active. I started reading about direct energy conversion devices, something which had interested me since the great gas shortage debacle in nineteen seventy-three. I found many of them in the literature; most are laboratory curiosities whose efficiency doesn't approach the current steam-powered rotary generators used by power plants today. Photovoltaic cells (solar cells) are the best of them, but so far their cost and energy efficiency isn't up to rivaling the heated-water / rotary method.

    The thermomagnetic converter, which almost became a mainstream reality in the nineteen fifties, caught my attention. Back then, this converter's efficiency was recognized as surpassing steam-driven rotary machines. However, the requirement of a fluctuating heat source made a working machine too complex and the idea faded from consideration.

    It occurred to me that with one small modification, the need for a fluctuating heat source could be eliminated—creating a small, homeowner-scale, device capable of turning sunlight *(or any other source of heat)* into electricity with an efficiency surpassing oil-powered power stations. But I kept thinking my modification seemed too simple—something so basic should have been either adopted or noted in the literature as impossible years ago. And since I mostly sat around anyway, I read book after book trying to decide if this idea had merit. Several years later all I could say for sure was that while my idea didn't contradict any physical principles, for some odd reason, no one had looked at it before. So, what started as the stay-at-home hobby I'd used just to keep myself occupied, took on more importance and I finally became convinced someone should check it out.

As it was difficult for me to leave the house for long periods, whenever the opportunity arose, I retreated to my shop to see if it might be possible to change the Curie Point of a ferromagnetic metal electrically. If it were doable, instead of being a thermomagnetic converter, it'd become a ferromagnetic generator.

Right now, the ferromagnetic generator remains in its infancy. While I have some strong indications showing how this effect can turn heat into electricity, I can't provide you with jump up and down, absolute, proof. *(I have sympathy for that guy who stood on a windy North Carolina beach and tried to convince skeptics that remaining off the ground for 120 feet would someday lead to aircraft capable of taking people anywhere in the world.)*

So yes, a lot more research lies ahead. Like anything of importance, the country will need to invest money, time and resources before getting something that can sit on your rooftop and power your house. That's why I'm publishing the underlying principles in the hope of getting research institutions to look into the idea. More than ninety percent of the ferromagnetic generator is <u>established, proven, technology</u>! This added insight suggests one day we can turn solar energy into grid-ready electric power with at least the same efficiency as today's coal, oil and gas powered generators.

The world needs to reduce its dependence on fossil fuel.

While the Earth's energy deposits are amazingly plentiful, they really are finite.

Oil, that some countries have it and all others need it, plays a great role in the global unrest that's costing so many lives. While a discussion of this tragedy is beyond my short book about an energy-source alternative, I should at least recognize the wars, terrorism, and brutal governments brought about by the world's overwhelming demand for black gold. People, our fellow human beings, are dying. All political spin aside, our world titers on the edge of catastrophe over oil. Even in oil-rich places where a hot war is not currently raging; unrest, instability, and disrupted lives are the norm and often encouraged by those who seek their resources.

Oil costs us far beyond its per-barrel price. Shipping disasters (seagoing tankers, pressurized pipelines and trucks), refining hazards, carcinogens released into the environment, and even the soul-destroying karma earned by those who, almost daily, compromise their integrity for some company's bottom line. The 2010 Gulf oil-blowout taught everyone a bitter lesson about the staggering cost of drilling for oil. Some research claims the true '2017' cost of a single gallon of gasoline would be well over fifteen dollars if all the costs, now underwritten by taxpayers, was included. And this estimate is thought to be much lower than for coal, who's long-term environmental and health expenses are too widespread to accurately calculate.

In any discussion of the cost of coal and oil, we should include the yet undetermined price of global warming. I argue the effects of destabilized weather patterns change the rules, and projects that appeared aesthetic only a few years ago are now economically viable—even necessary.

In just these last few years we've seen disaster after disaster caused by our over-revved weather system. Human tragedy aside, tornadoes, hurricanes, floods and fires cost us many billions. Their increasing frequency is painfully obvious to all without a closed mind. Much of the American West is drier than it's ever been in recorded history—including the years of the Oklahoma dust bowl! How much are the resulting fires and crop failures costing us? The heated atmosphere holds more water vapor which causes both drying and flooding, so as the west dries, record snowfalls and rainfalls send devastating floods and storms across the Midwest.

These tragedies are not confined to the U.S. We've all seen the reports of the fires and floods in other countries. Drought or flood related crop failures are becoming the norm. And of course, our melting Arctic ice caps—those great freshwater storehouses that, if released, will raise the oceans and flood coastal cities, even whole countries. Plus they could disrupt the flow of the ocean currents which keep our climate stabilized. Climate scientists, without a corporate agenda, agree the effects of global warming are about to get much worse.

By now, every thinking person knows when any form of carbon burns it creates the carbon dioxide which plants and algae removed from our atmosphere millions of years ago when the world was young and wild and hotter than today. I fear, after we've finished burning all our oil and coal reserves, atmospherically, we'll be back to the dawn of life on Earth. It won't be a friendly world for humanity.

Yeah, getting the ferromagnetic generator up to usable efficiency is going to take serious work. But we're spending so much money and risking our human future on our present oil / coal economy. If the country diverted less than one percent of last year's disaster cash into developing this idea, I believe there's a chance it can become the solution to Earth's energy problems.

* * *

Replacing fossil fuel with an Electro-Hydrogen economy.

Well, I assume you understand we're talking about hydrogen the chemical element. This has nothing to do with those bombs that were so terrifying during the cold war.

As an element, hydrogen is great stuff, raw material of the universe and all that. But it has few problems when used as a transportable fuel.

1. Nature's nearest "gas station" for uncombined hydrogen is the planet Jupiter. Most hydrogen on our little planet is chemically combined with either carbon or oxygen. It takes energy to yank it free.

2. Hydrogen is a gas. Not only that, but the Hindenburg disaster gave it a nasty reputation.

Still, hydrogen is <u>less</u> explosive than gasoline and most of the time we're handling that well enough. *(Did you see the film showing the Hindenburg disaster? Human tragedy aside, it runs for several minutes. If the airship had been filled with a vaporized-gasoline/air mixture, it would have run for about 0.2 seconds.)*

Running an electric current through water spiked with a acid breaks H_2O apart to give both hydrogen and oxygen as free gases. It's even efficient; each electron splits off a hydrogen atom. *(If you let the resulting two gases mix, it's called Browns Gas and can be used as a fuel. But it's easy enough to keep them separate and get free hydrogen and free oxygen.)*

Free hydrogen is a gas and the water is a liquid in this reaction. Going from a liquid to a gas requires energy. In this case it would be about a 15% loss. Nope, no way around it. So, turning water into hydrogen to use as a transportable fuel is going to cost us 15% off the top. For us, today, that means our method of creating electricity must be efficient enough to cover the nut and still create stored hydrogen energy cheaper than the equivalent amount of petroleum. Fortunately, solar energy is free!

Industrial realism:

A working ferromagnetic generator could easily exploit solar energy. It requires no boiling of intermediate fluids like water to turn the turbine—just focus the sun's rays into its core and once the working temperature is reached, electric current flows out. So, if this new style of generator could reach the efficiency of a thermal generator, could it run the country? Or asked another way, how efficient would a solar-to-electricity generator need to become to compete with burned-fuel power plants?

Photovoltaic—solar-cells convert 10 to 15% of solar energy to electric power depending on their time in service. I understand more efficient ones are available but not yet in common use.

Solar cells have not taken the country by storm even with all the tax and other incentives available to encourage their use. Part of this is because of their high price. They're mostly made of sand, so why do they cost so much?

Because sand needs melting in a furnace and then kept molten while being refined to seriously impressive levels of purity. It requires <u>burning much expensive oil or coal when making a solar-cell.</u> Once again, it's a matter of finding someone to be honest about the total energy and pollution cost in fabrication versus the energy produced in a typical cell's (realistic) lifetime. I'm thinking the current 15% Insolation to power ratio just doesn't make it except in specialized or taxpayer underwritten situations.

Don't give up on solar cells. A scientific breakthrough that upped their efficiency could happen any day. *(However, keep in mind the current price of solar cells is held down by those counties who burn vast amounts of polluting coal to refine the sand.)*

Let's move on: If 15% is too low, how about an energy conversion ratio the same as a coal or oil fired power-plant. That's around 40% depending on the age of the plant, and whether it's working at peak or below. *(The efficiency of burned-fuel plants is constrained by the Carnot Cycle. Normally, around two thirds of the energy stored in coal or oil goes up the smokestack or into the cooling water.)*

For a generator turning around 40% of the Insolation into electricity: *(Physicists like to call the energy in sunlight Insolation. Sometimes Physicists become envious of all those Latin names biologists get to use.)*

Some rough comparisons:

A square foot of sunlight in much of the United States gives around 17 watts. *(Watts have a time base. They are one joule in one second. You take your sunshine at the rate the sun gives it to you.)* 17 watts / sq. ft. is a crude estimate from the 5 watts available in northern winters to the over 40 watts land enjoying a southern summer receives.

Are you thinking about all those unused rooftops?

An average residential roof of 2000 sq. ft. at 17 watts / sq. ft. has 34000 watts when the sun shines. That's 34 KW/hour times 6 hours of "good" sunshine—which gives the homeowner 204 KW/H day or 6120 KW/H a month. Last April my Florida home used 2600 KW/H total, so I would need to turn 42% of the solar energy landing on my roof into electricity to get my electric bill to zero. *(Note: "Good" sunshine is when the sun is high in a cloudless sky. 6 hours is arbitrary, but seems about right.)*

With excess power in the daytime and none at night, I think running the meter backward and selling the excess power to the electric company to turn into hydrogen would be the best choice for a homeowner.

Motor vehicles and industry need more power than homes. We need to think bigger. A barrel of crude oil has approximately 6.3×10^9 joules. *(Joules have no time base. You can burn oil fast or slow.)*

Playing with the math. These are just my back-of-the-envelope calculations. There is no need to go into extreme detail in this book. More scientific studies have already addressed the subject in all the exquisite precision you can stand.

Land equivalent of a barrel of oil.

Most anyplace in America a square, 113 feet on one side, gives at least barrel of oil's energy every month. Even a square only 6.5 feet on a side produces a barrel of oil's energy every year.

Cloudiness plays a part, but most of the lower 48 states get around 2800 hours of sunlight each year with the desert areas getting more than 4000 (of the 40 watt a foot kind.) Given the 48 states have around 3.1×10^6 square miles; we're talking about getting the energy equivalent of 7,586 billion barrels of oil each year. *(Other (better) calculations are available from many sources; I'm told my estimates are low.)* With total conversion, we could meet our energy needs using just 0.15% of our land area. That's way, way less than what we've already covered with roads and buildings.

A typical nuclear reactor produces between 800 and 1000 megawatts of power and is expected to be on-line 80-90% of the time. *(If operated competently??—I live near Crystal River, Florida; enough said.)* However, solar is on-line around 32% of the time so the land needs to be increased X3 to equal a nuclear generator's output. 100% solar energy conversion isn't reasonable, so let's say our solar generator is as efficient as a modern thermal generator.

I'm coming up with a square mile of desert land receiving 1115 megawatts of solar power. Assuming a conservative 35% conversion, that's around 390 megawatts. Times three for offline at night boosts the needed land to 3 square miles. So a square, less than 2.5 miles on a side, would give the same output as a nuclear generator. Plus, no radioactive waste to store for a thousand years. *(Who wants to live within a few miles of a nuclear generator?)* With solar collection, there's no danger of an "opps" killing people and animal life and creating an empty zone like around Chernobyl or Fukushima. In fact, between the collectors, the normally overheated desert might become truck farm productive. A little clever planning and the land becomes multiuse.

To put this into perspective with a "thought" experiment. America's Death Valley covers around 1500 square miles. Piss off the environmentalists and condemn just ten percent for collecting solar energy. At 35% it would give us 58 Gigawatts of power. That's equal to 70 nuclear power plants! Or, or allowing for solar being off at night, around the full time equivalent of 30 nuclear plants. *(OK, I like Death Valley the way it is now. But there are unused bits, and much empty land in the neighborhood receiving the same level of solar radiation.)*

With that off-at-night constraint, any discussion of replacing coal and oil with solar must include hydrogen. While, individually, many people can collect their energy needs from their own rooftop and a few batteries, industry and transportation require an intermediate fuel. It's a nice coincidence one can be extracted from water and when used, will simply turn back into none-polluting water. *(With the added advantage that many current oil-dependent engines can be adapted or retrofitted to run on hydrogen, reducing the disruption of a changeover.)*

Besides describing my work on the generator, the other theme of this book is that developing alternative energy sources can provide a way out of the current economic and climate change troubles that threaten life on Earth. We spend many billions on less desperately needed projects, some serious money thrown at the foundation I'm presenting in this book, could change this generation's legacy from "The great failures" to "The great saviors."

This book is about our children's future.

CHAPTER 1

First, the good news:

Almost everything about the ferromagnetic generator is already in the scientific and historical literature. The reports I've found suggest Thomas Edison either built, or at least designed, one of the first models.

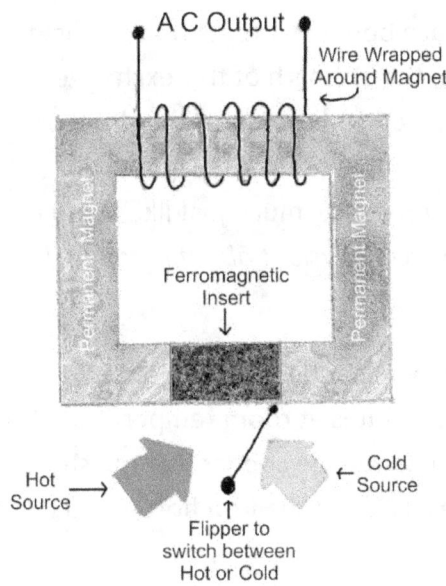

The bad news.

As originally conceived, Tom's generator needed a cyclic heat source. Heat needed to be pumped into the ferromagnetic core until exceeded its Curie Point. The core then became non-magnetic which caused a magnetic field to collapse and create a current surge in a copper wire. After the surge ended, the core then needed cooling until it was once again below its Curie Point—usually by dousing the core with a spray of cold water. That's my drawing of Edison's device.

Well, even Edison gave this idea up as being too slow and awkward.

But the fundamental idea has promise. All we have to do is eliminate the flipper thingy (cyclic heat source.)

How about, instead of making the heat in the core rise and fall, we find a way to shift the Curie Point a few degrees?

Then the heat (temperature) could remain constant, and we would let the Curie Point shift back and forth across it, alternately turning on and off the core's ferromagnetism. *(Note: there's a technical difference between heat and temperature, but we can use the words interchangeably in many cases.)* When the insert was ferromagnetic, a magnetic field would rise, and when it became non-ferromagnetic the field would collapse. This would be just the same as Edison's generator, except that we would have removed the wasteful and slow cooling bit.

I'm going to be talking about the Curie Point a lot. Here's a quick outline:

The Curie Point.

Ok, this was a neat discovery. Pierre Curie *(Marie's hubby)* found it in 1895. For any ferromagnetic material (5 elements and associated alloys) ((there's a sixth element, but it's one of those laboratory-provable only things.)) he discovered a temperature at which it suddenly stops being ferromagnetic. For iron, it's 766°C. Ferromagnetism is created by a special form of electron interaction called exchange coupling which takes place between atoms that are next to one another in the metallic crystal. At the Curie temperature, the strength of this exchange coupling equals the energy of the thermal agitation. Above the critical temperature, the atoms are wiggling around so fast the coupling comes apart and the sample's overall magnetism disappears. (It becomes paramagnetic, which is a fancy word for pretty much just like every else. *(FYI. Exchange coupling is a purely quantum effect. Ask a classical physicist about it and he'll send you to the next lab down the hallway.)*

For our generator, the thing about the Curie Point is that, being dependent on a quantum effect, it occurs rather suddenly. Hot 600°C iron is almost as magnetic as it is at room temperature. But a few degrees above its Curie point and magnetically, it might as well be copper. Let me show you a graph of how the magnetic properties of iron decrease as we get the iron hotter and hotter.

The B-H curve of Iron.

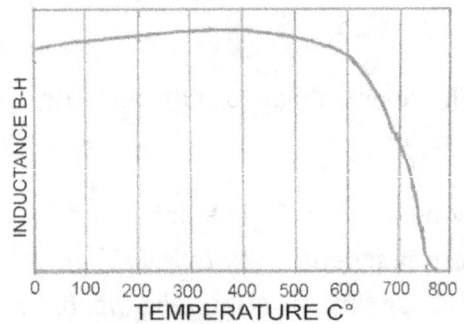

Most ferromagnetic substances have a similar curve, and let's not forget the BH properties are also dependent on the magnetic field strength the sample is experiencing. But for us normal folks, this is a good representation the way iron loses its magnetization as it gets hotter.

Along the bottom we plot the temperature of our iron sample; from ice-cold all the way to cherry-red heat. The vertical scale is permittivity (how many lines of flux will go through the iron rather than air.) So, the bottom represents air. Normally, the magnetic force prefers to ease on through transformer iron around 7000 times more than air. Where you would have only one lonely flux line in the air, offer it a bit of iron and you'll suddenly have 6999 of its buddies crashing your party. That's the point where the line touches the left side of the graph.

So: looking at the graph we see that around 770°C our hunk of iron has lost its magnetic properties and to an external magnetic field it isn't any different from the air. But look at the last little bit. See how steep the curve is? Keep that in mind.

Every ferromagnetic material has a unique Curie Point. Different alloys, different purities, even different annealing processes, all result in a different point.

Curie Points of pure elements range from Gadolinium @ -6°C, Nickel @ 358°C, Iron @ 770°C and Cobalt @ 1130°C. Various alloys of these metals and other alloys exist that run the full temperature gamut.

This special point, where a ferromagnetic material becomes nonmagnetic, is subject to a wide variety of restrictions. Still, for a given sample, researches have not yet been able to change the break-down point more than a few degrees. Usually by applying great pressures or strong magnetic fields or electric currents.

But it does change. And if it changes a little, can we find a way to make it change a bit more?

Looking at the graph we see the change from magnetic to nonmagnetic is not instantaneous, but looks like a gently sloping meadow just before a steep, but not vertical, cliff. I'm convinced a ferromagnetic generator need not cycle the ferromagnetic material through its Curie Point, but would work best in those last few degrees

Magnetic Lines of Force:

They're imaginary, but work as a way to talk about the not-well-understood force that's at the foundation of our universe. The term came about because if you put a piece of paper over a magnet and then sprinkle it with iron filings, the filings from a pattern that looks like lines going from one of the magnet's poles to its other. Unlike the electric force, a magnet's force lines are closed and always travel from the north end of a magnetic dipole to its south end.

Like everything else in nature, lines of force seek out their lowest energy state. That means they want to be as short as possible. In addition, they repel one another: a great deal in air, not so much in a ferromagnetic material.

Basic magnets:

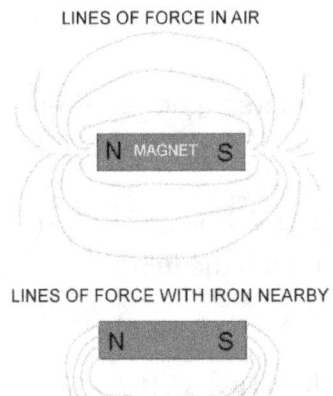

Permittivity is a relative term, used to show how much more magnetic flux will travel through a sample compared to air.

So, as AIR = 1 the higher numbers denote a greater ability to carry flux. Iron at 7000 will carry seven thousand times more lines of force than the same physical space if it contained air. Like the Curie Point, different metals and alloys have different Permittivities. Some examples: Cobalt = 250, Nickel = 600, Mild steel = 2000, Iron = 5000, Silicon iron (Transformer iron) = 7000, Permalloy = 100,000.

And top of the class, Supermalloy = 1,000,000.

More basic stuff we should review so this Generator makes sense:

Magnetic Domains: (Or why every hunk of iron isn't a mega-magnet.)

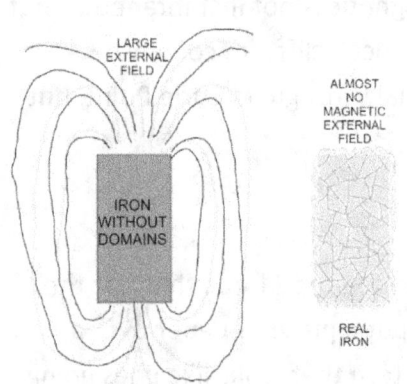

Classically, all electrons spin like a top. As they have an electric charge (negative in our universe) this spinning gives rise to a magnetic field.

<u>Basic rule—a moving electric charges create a magnetic field.</u>

Let's call our spinning electron and its magnetic field a dipole. Although it's unbelievably small, it acts just like a bar magnet and has a North and South Pole. Still, you don't get much magnetism from a single electron. But in ferromagnetic metals (and their associated alloys) some special electrons link in such a way that their elementary dipoles lie parallel to one another. And that lets their magnetism add. If you could get every dipole in a hunk of iron linked in harmony with all the other dipoles in the iron, it would create one of those ridiculously-powerful cartoon magnets.

(In most materials, two electrons pair up head-to-toe so their magnetic fields cancel each other out. An oddity in the size and the way the orbitals mesh in a few metals causes them to allow an unpaired electron to dangle its magnetic field out into space.)

Now remember, dipoles align with one another because doing so represents a lower energy state. For these little guys, it's like the difference between standing on the side of a really steep hill or sitting in a lawn chair down at the bottom. They really want to line up with their buddies. And they do. Each tiny dipole adding to the whole so the magnetic field becomes stronger and stronger, causing the magnetic lines of force they emit to start getting larger, more concentrated and, because they continue to repel one another, extending farther and farther out beyond the iron bar.

Lines of flux (from linked dipoles) cannot cross. Plus, they must always loop from the individual dipole's North Pole all the way to its South. As more and more dipoles link and add their magnetic flux together, the external magnetism increases. So as the dipoles link, the associated flux spreads further and further out into the air (space) around the iron.

Also, those lines of flux *(the lines I've drawn)* are always seeking to be as short as possible. At some point, the need for the dipoles to link is overcome by the power of the lines of flux trying to be both as short as possible and not crowded together with other lines.

A compromise is reached. **Domains**: tiny places within the iron where all the dipoles align but are not so large their individual lines of flux need to travel very far from their North Pole to their South. See the picture with all the little cells—in real life they're about a tenth of a millimeter across.

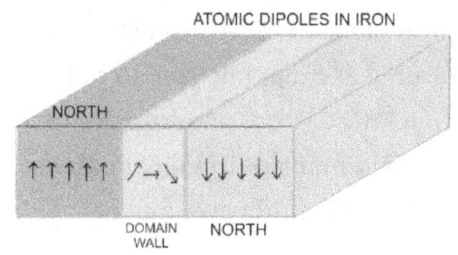

By dividing itself up into millions of tiny subdivisions, the iron keeps the length of its lines of flux as short as possible but still allows the dipoles to link. *(And for the dipoles this is like Friday night at the beer garden.)*

But: what about adjacent domains? Here we have a bunch of dipoles all happily lined up facing one way and right next door a bunch that doggedly face some other way. So, between them a wall forms—a place, maybe 1000 atomic diameters wide—where the linking of the dipoles weakens so they can gradually rotate from the orientation of one domain to the next.

Domain Walls can move. *(The atoms stay in place and the wall moves like a football crowd doing the Wave.)*

Something to remember: The domain walls are smallest at room temperature and expand as the temperature rises.

Should our iron sample find itself in an external magnetic field, those domains whose spin just happens to lie in the direction (or almost in the direction) of the invading field will grow by pushing their walls into the territory of less lucky domains. As this happens, our once (overall) nonmagnetic iron bar starts to take on magnetic properties. As the external field increases, the favored domains grow.

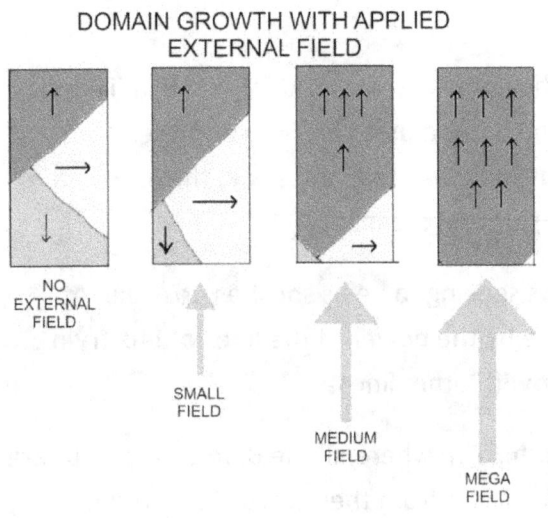

Eventually, the iron reaches a point where all the almost-aligned domains are as big and fat and in compliance with the external field as they can get. After that, any increase in the overall magnetism of our sample occurs as almost-aligned-but-not-quite domains give up the fight and flip to completely align themselves with the external field.

When this effect is plotted on a graph, a pronounced "knee" shows up, as the easy task of moving walls ends and the harder job of flipping the stubborn hold-outs begins.

Take away the external field and the iron tries to return to the state it was in before you went and messed with it. This is where the terms "Soft" and "Hard" iron arise. In "Soft" iron the tiny domains quickly reappear, and little prevents the walls from forming and zipping back to their original positions, so, externally, the iron stops being a magnet. In "Hard" iron, non-iron atoms and crystal deformities prevent the walls from slip-sliding around and many of the tiny domains remain mostly aligned in the direction of the field that turned them. This is how we make permanent magnets. *(Well mostly)*

As the reason the ferromagnetic generator works has a lot to do with electric motor and transformer theory, we should review their basic principles.

First, Electric motors.

Reduced to its basics: When electrons move inside a magnetic field it creates a vector force. If those electrons are in a wire, the force is transferred to the wire and it will, or at least try to, move. Many, many physics books, engineering books and websites will explain the details of this in all the depth you can stand, so let's take it as a given.

One detail to note: To create an electric motor, either the electrons or the magnetic field has to be moving. Normally, that's electrons traveling from negative to positive. However, equally effective is a changing magnetic field; one that is increasing or decreasing. (Yes, some motors employ both methods). This is a basic Maxwellian given: a moving electron creates a magnetic field and a magnetic field that is changing causes electrons to move.

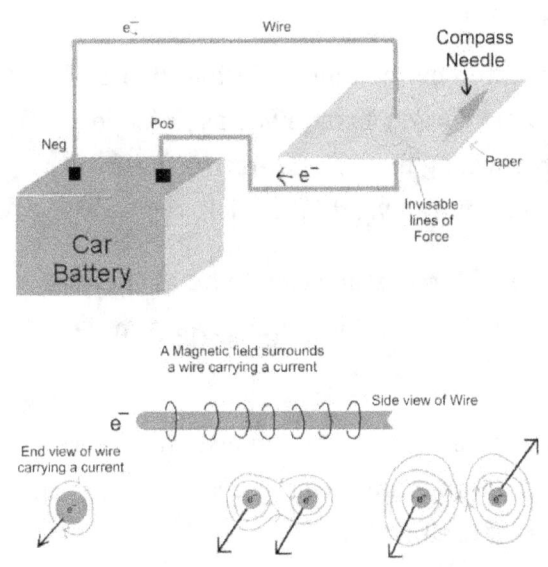

Here is a simple diagram of the magnetic field that surrounds a wire carrying a current. A compass needle will reveal the invisible magnetic field.

Err, I forgot to show a resistor in my picture of a wire connecting the two poles of a car battery. Shorted car batteries can produce over 200 amps. *(You really, really don't want to do that without a resistor.)*

Note that when the electrons in two different wires are traveling in the same direction, their magnetic lines of flux tend to suck the wires together. But when the electrons are traveling in opposite directions the wires are pushed apart.

To create a vector force, change is fundamental. Either electrons have to move or the magnetic field has to either move or change. (Change = The field grows stronger or weaker.)

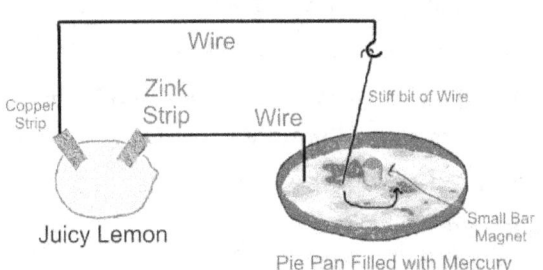

While we're talking about electric motors, let me include this diagram of the <u>very first one</u>. When the two differing metals are inserted into the lemon, the stiff wire rotates around the little permanent magnet sitting in the pan of Mercury.

(The vapors from open mercury are poisonous and cause mental problems.) ((Like Lewis Carroll's Mad Hatter—Years ago Hatters used mercury to make felt.))

This is grandfather of all modern electric motors. The great engines that today power ships, submarines, and 100-car locomotives besides doing the more important job of rolling down the windows in your car, grew from this humble laboratory toy. Perhaps the Ferromagnetic

Generator isn't much right now, but don't forget, like electric motors, everyone starts out small.

We need to talk about magnets as they make electric motors turn.

First there are the hunks of iron that stick to, or pick up, other bits of iron. They have two poles, called North and South. We can go on all day about the neat tricks you can do with permanent magnets.

You need to be aware how the invisible lines of force created by the dipoles inside the iron live by two rules: be short as possible and shove all other lines of force as far away as possible. However, you remember how much those lines prefer traveling through a ferromagnetic material. If you want to get the lines closer together, a bit of iron will do the trick.

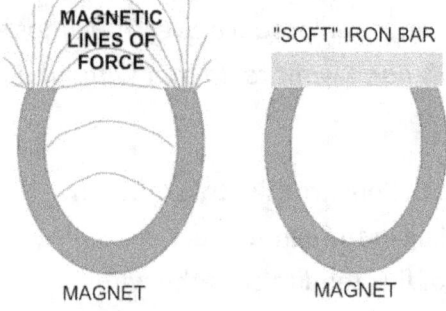

Note, it's the lines of force, still trying to be as short as they can be, that suck the soft iron bar toward the magnet.

Well, I'm sure you're all aware how unalike magnetic poles attract one another and like poles will shove each other's backside off the barstool.

Second are the electromagnets.

When an electric current travels through a wire, it creates a magnetic field around the wire. The greater the current the stronger the field.

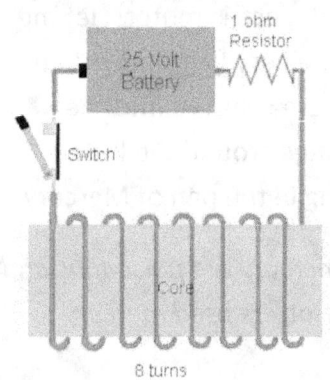

When the current flowing through two wires is in the same direction, the magnetic lines of force the moving electrons create add together. If we wind the wire around a bobbin, all those turns create an even stronger magnetic field. This is how electric motors, the kind that don't use permanent magnets, produce a field able to do useful work.

This results in one of those easy equations: F=NI. (Magnetic force is equal to the turns multiplied by the current.)

When the battery is connected, current flows through the wire and the magnetic flux produced by each individual turn of wire adds to the flux created by the other turns. For clarity, I'm only showing a few turns of wire from what would normally be several hundred.

Bobbins can be wound with an iron core to concentrate the flux. This is how an electromagnet is created. Note that an electromagnet only possess a magnetic field when the battery sends current through the wire.

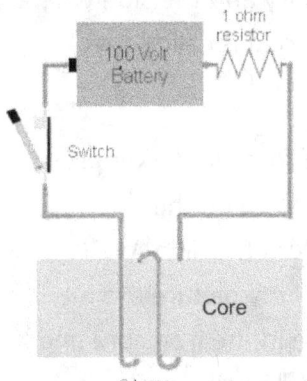

As the magnetic force created is equal to the number of turns times the current in the wire, you can build a magnet with many turns and use just a little electric current. *(Like a doorbell with a bobbin of super-fine wire.)*

The second picture differs in that we have fewer turns but more current flowing from a more powerful battery.

We're creating the same magnetic force using just a couple of turns with a whopping great current running through the wire.

The formula is simple: current times turns equals FI (field strength)

(Note: I used a one ohm resistor in those pictures. That allows the current (in Amps) to be equal to the battery voltage. (In Volts) It makes the example kinda neat and easy to understand, just don't go and calculate how downright hot those resistors are going to get if anyone is foolish enough to build this circuit.)

For DC circuits, Power (in Watts) is equal to the amps squared times the resistance.

And now (Electric) Transformers.

Transformers use a changing magnetic flux to either boost up or reduce a power line's voltage. For the same amount of power, low voltages require high currents. High currents heat up their wires and waste energy. Power companies use transformers to create very high voltages which they send for miles and miles without much loss. (Because the current is so low) Then, at the other end of the line use reducing transformers to change the power back to safe-to-use-in-the-home voltages. So when you plug in your toaster, you're not toast.

Remember, a moving electric charge creates a magnetic field. (Lines of force) Naturally, the opposite is also true. A changing magnetic field will cause electrons to move.

Here's my rendition of a simplified transformer. Normally, the two (or more) windings are wrapped one on top of the other, but pictured this way makes it easier to see what's happening.

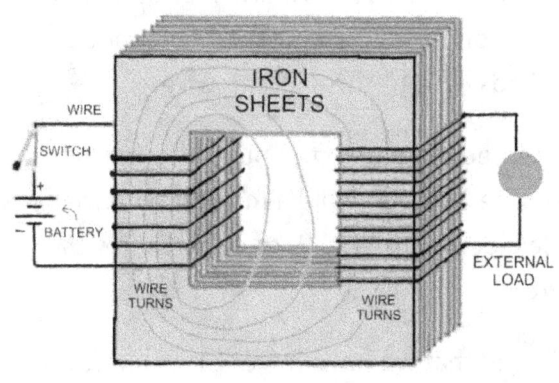

Note that the side (windings) where the power comes in is called the Primary and the side where the power goes out to the load is called the Secondary.

I drew the invisible lines of force in darker grey. The large square is supposed to be the iron transformer looking at it from the side. Well, some of you noticed *(once again)* my picture is a bit wrong. I show lines of force in the iron core and I depict the switch in the "off" position. That's artistic license, that is.

Besides, we can pretend it's only been a millisecond since the switch was thrown and the lines of force have not yet completely collapsed. So in this picture imagine those lines are in the process of zipping back to the Primary side.

It helps to think of lines of force as similar to a rubber band. They have no beginning or end and must always travel from the north end of their dipole to the south end. They can be little or big. But they cannot be cut so are always in a loop.

(Big as in those magnetic lines of force which emerge from our Earth at the North Pole and travel deep into space before returning to the South Pole. This fortunate bit of physics protects life from the charged particles in the solar wind.) ((Magnetism keeps you alive.))

Back to the picture. As the lines are created by the current going through the primary side wire, in order to travel all the way around the iron core (which they really want to do) they must zip across the air-space in the middle. And as they zip, they must pass through all those secondary windings. Remember the basic, and really convenient, law of our universe: A moving magnetic field (changing) will create a current in any wire it passes through.

It gets better. The flux created by one turn creates *(more or less)* the same voltage in any similar length of wire it passes through. So, let's pretend that each turn of the wire wrapped around the primary side creates one line of flux. Now, that line must zip across the air space and pass through the wires wrapped around the other leg of the transformer. If, for every turn of wire on the primary side, we wrap two turns of wire on the secondary side, that flux will need to pass through <u>both</u> secondary turns to get into the iron where it wants to be. Each turn gives the same voltage as the voltage on the primary, so overall; at the secondary's output we get a voltage of twice that of the primary. (If we want to talk about less turns on the secondary side, it becomes a bit more complicated. Then we need to start talking about energy transfer as less turns means less voltage but higher amperages than on the input side are available.)

In the picture I've drawn a switch. And now you know the only time there will be a voltage on the secondary is during the split second just after it's turned on and off when the lines of force are zipping through the secondary wires. (The light flashes.) But when the switch is on or off for more than what scientists call five time constants and we call the blink of an eye, there's no power coming out of the transformer. This is where AC power comes in.

Real transformers are AC only. If you connect the primary side of a transformer to a DC source like a battery, nothing stops the current from rising to the maximum allowed by the wire's low resistance. If a fuse doesn't blow, it will overheat and catch fire.

AC is a sinusoidal current which goes alternately in one direction, and then in the other. Sixty times a second in the United States, 50 times a second in England. I'm not sure about any other places. It's set by the number of poles in the rotating generator and how fast the engineers have the bundles of wire spinning around. That means in a transformer connected to an AC line, the magnetic lines of force are always either expanding or collapsing and the energy transfer in a transformer is continuous.

Well, there's a lot more to be said about transformers and AC current, but this isn't a textbook so I'll just go over what's necessary as it comes up.

Reviewing passive components.

There are things used in electric circuits we should talk about. Science-types give the name passive components to the bits which consist of only of an in-wire and an out-wire. *(The word wire can be misleading. Photocells and such are not passive as light or radio waves act as a third input.)*

I've talked about resistors already. They're the things that impede the flow of electrons by changing some of your hard-won electric power into heat. While that doesn't sound like the most sensible idea, in almost every electric circuit, resistors are the most commonly used component. Plus, as resistance stops our world from ending in one spectacular flash—they have their uses.

In building our generator, we'll need a better understanding of the two other passive components:

Inductors and Capacitors.

Ok, imagine you're the supervisor responsible for a trainload of cattle delivered to the Chicago stockyards. You have ten thousand steers in a holding pen with two chutes leading to the slaughterhouse. When you open a holding pen gate, the steers, thinking it's a way to return to the Texas grasslands, thunder down whatever chute becomes available.

The first chute is labeled "Inductor." You open the holding pen gate and cattle go charging down the chute until they encounter a second gate. It's closed. But as soon as the first steer touches that gate, it begins to swing open. Soon, one steer, then another, then two at a time can pass the slowly opening gate. After a short delay, the gate is fully open and the other steers rumble past as if it wasn't there.

You still have many cattle, so you open the gate to the chute marked "Capacitor." Once again, the steers charge down the perceived escape route. The lead steers don't even notice the fully open gate half-way down the chute. But as soon as the first one passes that gate, it slowly begins swinging shut. As the gate swings against the flow, the pre-hamburger is slowed and, when the gate is fully closed, those steers which haven't passed the gate remain trapped.

Ok, stop thinking about all those juicy steaks and come back to the less-edible world of electronic components. Inductors and capacitors work as the mirror of each other. Inductors, at first, resist the flow of electrons and after a delay *(yeah, five time constants)* pass current freely. Capacitors do the opposite. First passing the initial current and after that same time delay, stopping it completely.

For us, considering the principles behind the ferromagnetic generator, we need to think about the special time when the "inductor chute" gate is swinging open. We make an inductor by winding wire around a bobbin. We know around every wire carrying a current there exists a small magnetic field. Winding wire around and around a bobbin makes each turn's lines of force add as they are all going in the same direction. But remember how those lines of force push each other away. Those lines hate to be crowded, so as the current increases, the magnet field is pushed out into space, absorbing more and more energy from the electric current as the field becomes larger and gathers strength.

Note: both inductors and capacitors <u>store</u> electric energy: Inductors in their magnetic field, capacitors in an electric field. They can both give it back later. I have personal experience in how painful *(and expensive)* that can be.

When we apply a voltage to an inductor, at first, no current flows. Consequently, there is no magnetic field. As the current begins flowing through the coils, the magnetic field grows proportionally. After five time constants, the magnetic field is as large as it can get and only limited by the small resistance of its copper wire.

During this time (while the magnetic field is expanding) other forces pop into existence. Remember a changing magnetic field creates an electric field. Inside the magnetic field it's as if there exist billions of tiny capacitors, each one seeking to orient nearby electrons to its own polarity. That means they're exerting a force on the electrons, one that will counter the thermal agitation of the atoms they orbit.

<u>Generated electric field created by changing Magnetic field.</u>

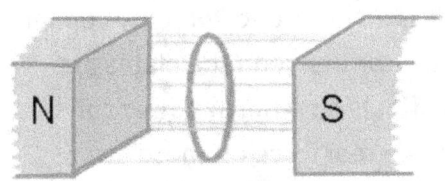

This is a close-up of an electromagnet's two poles in that <u>short time</u> when the magnetic field between them is increasing. As the lines of flux become more dense, a circular electric field *(the circle as seen from the side)* appears. Any electrons inside that field will try to align themselves with it. *(In addition to the dipoles trying to align themselves with the external field.)*

Err; don't let my artwork deceive you. The electric field occupies the <u>entire area</u> where the magnetic field is expanding. I figured a big smear wouldn't get the idea across.

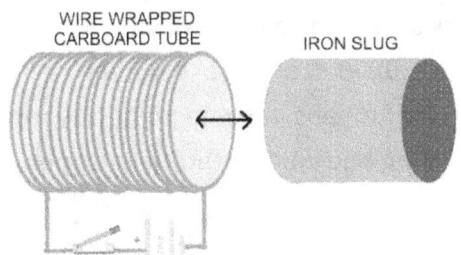

This is a simple Inductor. It's just insulated wire wrapped around a core. Shove a bit of iron into the core and you've upgraded to an electromagnet. (When a current is flowing through the wire, the hollow core will pull the iron into itself.)

As we know, the iron increases the lines of flux, which means more power is stored in the magnetic field. *(FYI—Inductors without iron cores work best at high frequencies—add some iron and you get lower frequency but more power stored in the field.)*

Now a word about Capacitors.

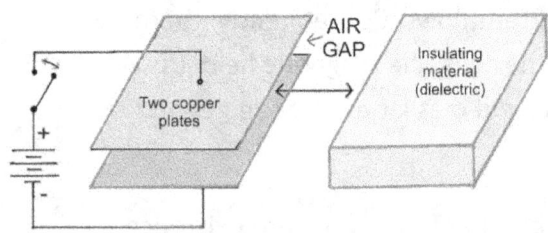

In a capacitor, the battery wires connect two conductive (like copper) plates that are not quite touching so the electrons can't jump the gap between them. Current flows until the voltage created by the excess electrons piling up on one plate equals the battery voltage. The air between the plates acts as a decent insulator, but there are better. (Like polystyrene.) If, after the current has stopped flowing, you insert a better insulator between the two plates, current will once again flow. (For a short time.)

(And, just like the inductor grabbing the iron core, the plates will snatch that hunk of dielectric out of your fingers.)

I thought you should know this so you don't think the dielectric is just there to keep the copper plates from touching each other. The electric dipoles in the insulator between the plates are distorted by the electric field and the extent of that distortion affects how much power can be stored inside the capacitor. Keep in mind any electric field exerts a real force on those atoms inside it.

It would take books and books to examine all the tricks you can do with inductors and capacitors along with all the neat ways they mirror each other. But these are the highlights as far as the Ferromagnetic Generator is concerned.

Oh, one other detail: a transformer is essentially two inductors arranged so one can pass its magnetic field to the other. You have a transformer somewhere outside your house. All the electric power you're using is passing through that device, and that means at some point all your power existed as magnetic lines of force expanding and contracting and cutting through copper wires.

There's a lot of power in a magnetic field.

CHAPTER 2

Can we build an electrical generator by manipulating the Curie Point in magnetic materials?

Remember Thomas Edison's proposed generator. In fact, the scientific literature contains a number of similar generators.

In the *Journal of Applied Physics* (Volume 30 Number 11) J.F. Elliot describes a device he calls a Thermomagnetic Generator. Essentially, I think he summarizes and works out the Carnot efficiency of the designs drawn by Thomas Edison and a number of other researchers. Back in 1959, when he wrote his paper, there was some speculation such a generator might work on satellites, as when they rotate in the sunlight they are alternately heated and cooled. In time, photovoltaics proved to be simpler and more reliable.

I've drawn a <u>non</u>-Curie Point generator which illustrates the basic concept of how changing permittivity generates power in Thomas Edison's device.

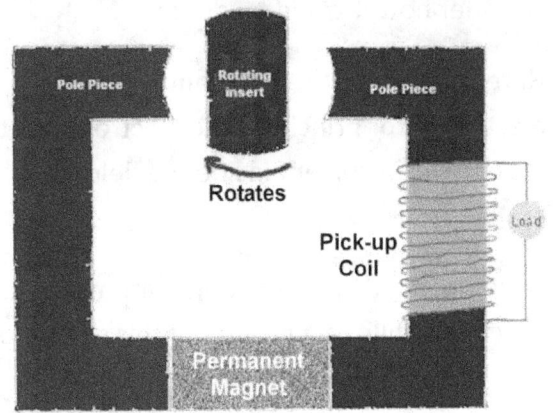

Instead of a heated and cooled insert, picture a rotating piece of ferromagnetic material between the poles of a permanent magnet.

In this picture when the inset is horizontal, the magnetic circuit is complete and when it's vertical (like shown) we have a substantial air gap which causes the total flux to drop.

At every half-rotation, the rising and falling magnetic flux gives the pick-up coil a burst of electric power.

The power for this generator comes from whatever source you're using to rotate the insert. If the output was connected to a light bulb, and you were cranking that sucker around by hand, you'd find it hard going as your muscle power would make the bulb glow.

In the ferromagnetic generator, we replace the rotating bit with an insert that doesn't move but changes its permittivity with temperature. Hot being less, cool being way more. In all those early devices, they heated the inset material past its Curie Point and then used a cooling system *(Old Tom with his bucket of water?)* to re-establish the magnetic field. This need for a constant rapid temperature change has always rendered this method of electric power generation an inefficient textbook curiosity.

I propose removing the need for real temperature fluctuation, so a steady heat input can be changed into electric power. The heat can come from any convenient source: Gas, oil, coal, nuclear or solar. It's only necessary to raise the reactor core to just below its Curie Point temperature. *(Remember that steep slope at the end of the B-H curve.)*

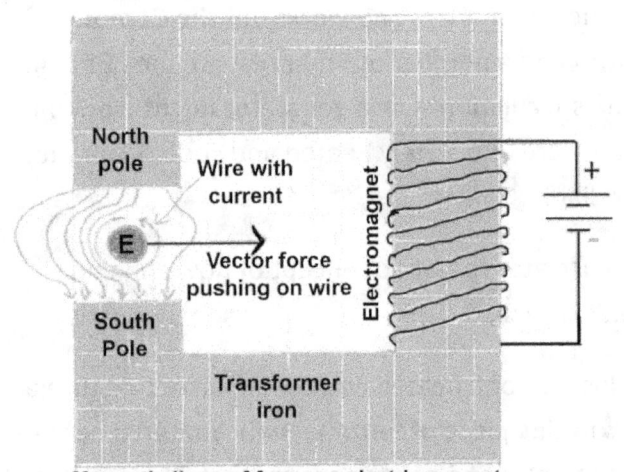

Magnetic lines of force on electric current

The most logical thing is to take advantage of the sudden decrease in permittivity just before the rising temperature causes a ferromagnetic sample to become non-magnetic. In effect, we electrically shift the Curie Point up and down by 50 or 75 degrees and take advantage of the great change in permittivity it causes. Let's consider how it could be done:

Here, I've drawn the basic outline of the electric motor principle. The electromagnet creates a magnetic field between the pole pieces. A wire carrying a current inside the field experiences a Vector force. That is, it tries to move.

We need to be aware the force is actually on the electrons circling the wire's copper atoms. Even if the wire is physically restrained from moving, those electrons still feel pressure. If this sounds a bit iffy to you, check out the Hall Effect, it works because of this principle.

I think a new kind of generator could work almost the same way. Instead of a copper wire, I'm going to substitute iron. I'll heat it almost to its Curie temperature, but not above that point. Unlike the wires in a motor, the iron will remain locked so it's unable to move. That means, all the force will be transferred to the electrons making up the iron's atomic dipoles. With any luck, the dual force of the electric and magnetic field will overcome the heat's thermal agitation and hold the dipoles in the position they would assume at a lower temperature. The result is (overall) the iron might pass the magnetic flux of the electromagnet as if it (the iron) was actually at a lower temperature. Externally, it will appear as if the iron's Curie Point has shifted.

Obviously, when the current stops flowing through the iron sample, the force on the dipoles ends. The iron then reverts to the low permittivity associated with its true temperature. That is, all the extra magnetic flux it's carrying gets rejected. As the collapsing magnetic lines of force cut through any copper wires in their path, they'll generate an electric current.

Here's how a motor would be modified to make it into a Ferromagnetic Generator.

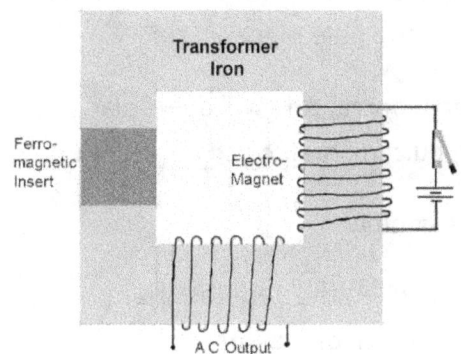

Note: The copper wire has been replaced with a hunk of iron. Plus, I've added a second winding to extract power from the expanding and collapsing magnetic field.

It's kinda like Edison's generator isn't it? Think about the ferromagnetic *(iron)* insert bit. When it's hot, almost at its Curie temperature, it has low permittivity. That is, it can pass very few of the magnetic lines of force the electromagnet is generating. But, when a current, at right angles to the magnetic field, passes through the iron—it can accept many more lines of flux. So, those little guys will take the easy path all the way around the transformer-like circuit.

Then we kill the insert current. Suddenly the circuit goes AGGH! Because the hot insert can't support all those lines of flux. They'll collapse. And as they collapse, they must pass through the output coil and causes electrons to flow inside that coil's wire. Moving electrons takes energy, and just as in Edison's generator, the only source of the energy is the heat inside the hot insert.

So, every time the current in the core stops or starts the flux from the electromagnet either expands or contracts. And each time it does, power (AC power) comes out of the pick-up coil.

Here are the mechanical bits I believe are necessary to make a Ferromagnetic Generator:

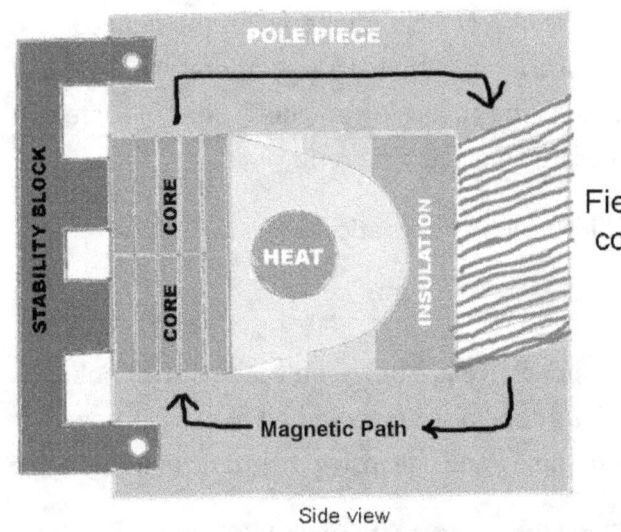
Side view

1. A coil of wire to make an electromagnet. (The pick-up coil can be wound on top of it.)

2. Pole pieces. These transfer the magnetic flux to the core.

3. The core. Made from a ferromagnetic material, preferably one with a lower Curie Point than the pole pieces.

4. A heat source. Any will do—Solar, nuclear, burning fuel.

5. Insulation. This device has hot bits and bits that need to be cooler.

6. Stability block. I put that in the diagram so everyone's aware when the generator is in operation, there's a vector force pushing the core. Something, a block or well-braced construction, has to prevent it from moving or vibrating.

7. Not shown:

The core has a changing electric current flowing through it. In this side view, the electrons would be coming out of the picture toward you. To keep the amps down to a reasonable level, it pays to segment the core metal.

Well, this all sounds too simple to be true, doesn't it? You're right. As I've described the generator so far, it doesn't work. *(Tried it, didn't I.)* I suppose if it did, someone would have discovered it long before this.

Onward to the tricky bits.

CHAPTER 3

THE DETAILS

Now you're going to understand why I kept going on and on about changing magnetic and electric fields and how one creates the other.

An electric current traveling at right angles to a magnetic field puts a vector force on the wire, but it doesn't stabilize the agitated magnetic dipoles in a hot ferromagnetic sample. Nor does an external magnetic field alone stabilize the dipoles. I'm convinced the only time this can happen is when both the electric field and the magnetic field are changing. For that short beat, of maybe three or four time constants, when both the electric field and the magnetic field are growing or collapsing, the magnetic dipoles should experience a combined stabilizing force against thermal agitation.

Think of it this way. A rising temperature causes enough thermal agitation to break the electron exchange coupling which creates ferromagnetism. Just before the Curie Point, the atoms are vibrating enough so they are alternately aligned sufficiently to couple and then disturbed enough so their coupling breaks. The domain walls have expanded to many times their normal size.

In the presence of four external stabilizing fields, I believe any residual exchange coupling will be reinforced.

These pictures are my crude representation of what is actually a <u>quantum effect</u>, so understand they're not especially meaningful. *(Many quantum effects manifest classically at the macro level.)* But they do give an insight into why a changing magnetic and electric field might momentarily cause a ferromagnetic sample to increase its effective permittivity. And why this condition will only take place in that hot-but-not-above-the-Curie Point zone.

Things to note:

It's the unfilled <u>3d</u> electron shell in an iron atom that undergoes exchange coupling with its neighbor to create a magnetic dipole. But it's the electrons in outer <u>4s</u> shell that are considered "free" and responsible for the metal's electrical conductivity. <u>Not the same guys.</u>

Quantum theory suggests that while thermal agitation (at high temperatures) can give the dipoles any orientation to the external field, only a limited number of possible orientations will reinforce the exchange coupling. That is, you're not going to get complete cooperation so the sample cannot return all the way to its room temperature permittivity. *(Bummer that.)*

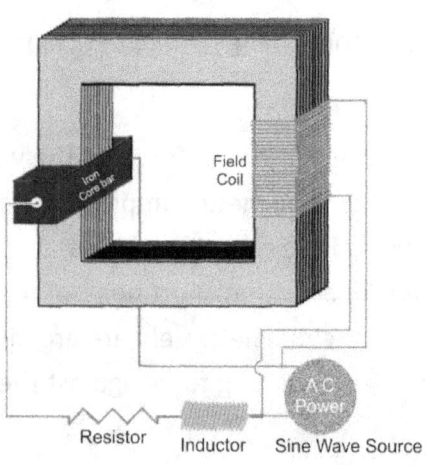

To better picture this, I've drawn the basic electronics necessary to establish a changing electric and magnetic field going through the portion of an iron bar that's between the poles of an electromagnet. Remember, the conditions are only correct during the short interval when the magnetic field and the electric current are both either expanding or contracting.

Unlike the copper windings wound around the iron of the electromagnet, the iron bar has next to no resistance or inductance. Phasing is important. The changing electric field inside the bar has to perfectly match the changing magnetic field created by the electromagnet. To do this, I'm showing external passive components in the wire leading to the iron bar. The resistor matches the wire's resistance and the external inductor matches the electromagnet's inductive delay. That way, the electrons coming out of the AC power source encounter two identical paths.

We're going to focus on that section of the iron bar between the poles of the electromagnet. But first let me draw a <u>classical</u> interpretation of an atomic dipole. In a classical sense, we can imagine them as similar to those toy gyroscopes you've seen kids play with. It's the spinning of the electrons, which carry an electric charge, that creates the magnetism. Err—all electrons spin. But normally they pair up head to toe so their spins cancel each other out magnetically. *(That represents a lower, and preferred, energy state.)*

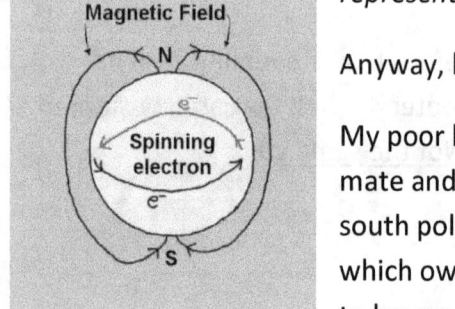

Anyway, here's my rendition of a single magnetic dipole.

My poor lone electron spins *(inside direction lines)* around without a mate and that creates a magnetic field giving the dipole a north and a south pole. *(Outside magnetic field lines)* The nucleus of the iron atom, which owns this electron, when heat-agitated, causes the whole thing to bounce and twist around.

With me so far?

Now, just after the switch has been thrown, between the electromagnet's poles, quite a bit is going on. First, "free: 4s" electrons in the iron begin moving in response to the electromotive force of the battery. Nothing unknown about that, it's what happens when you switch on a flashlight.

Second, magnetic lines of force, created by the field coil, surge around the low permittivity iron path and pass through the iron bar. When these magnetic lines encounter the moving electrons, they interact and create a vector force that wants to physically move the iron bar. We'll assume the bar is locked in place and can't move, but the force still distorts the orbitals of the iron's electrons. Again, nothing unknown, that's the basic motor reaction.

But, for those few time constants just after the switch has been thrown, two other forces flash into existence. The changing magnetic field creates, inside the iron bar, a <u>transient</u> electric field. And the changing electric field coming from the battery creates its own <u>momentary</u> magnetic field.

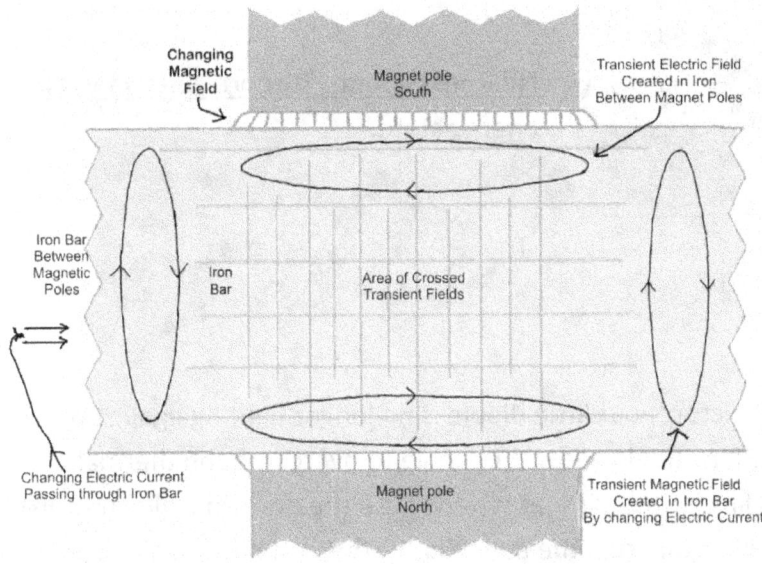

Here's how the two created fields cross. (Remember, these are <u>not</u> the electric and magnetic fields created directly by the battery—they're still there, but I'm not going to include them and make the picture so confusing you run out of the room.)

OK, drawing invisible forces is kinda tricky. The ovals represent circles viewed from the side. The fields they create extends through the iron bar. I hope you get the idea.

The next step is to put our little dipole, as a representative of a few hundred trillion of his buddies, into the picture:

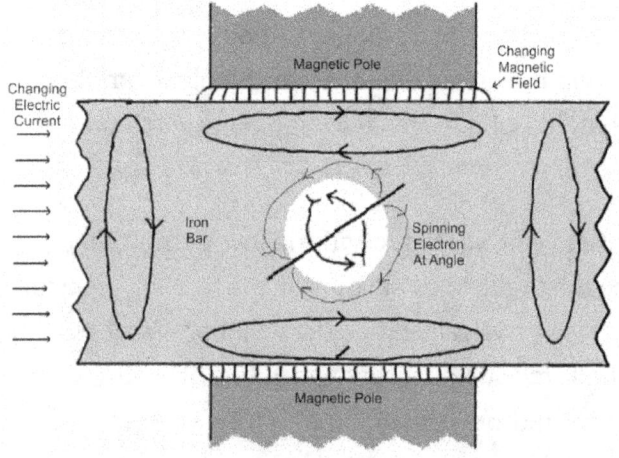

First, I'm showing a picture the dipole bouncing around so it's at an angle to the <u>momentary</u> created fields:

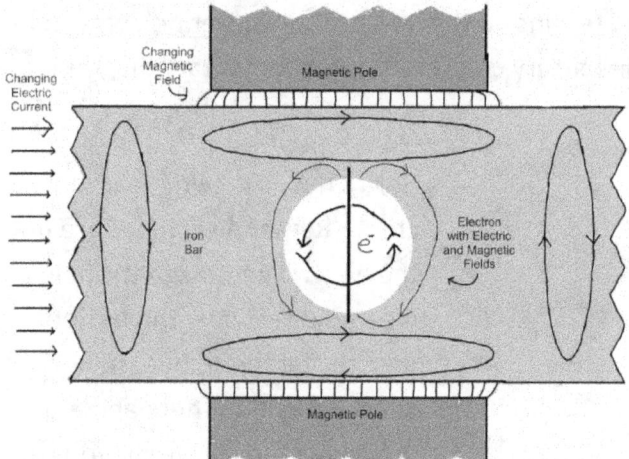

And then see how much neater everything looks when the dipole lines up with the fields.

<u>Neatness counts. It represents a lower energy state.</u>

I think you can see how in this second picture, our little dipole is no longer fighting against those external fields. This preferential condition should allow the dipole to overcome the thermal agitation. Then the lonely dipoles will link with their mates and cause the domain walls to shrink. Microscopic domains reappear and once again rule the day. That is, the iron becomes more magnetic.

Maybe I didn't explain this over-well. Yes, applying a magnetic field to a ferromagnetic sample above its Curie Point does cause the dipoles to align with that field. *(As best they can as they're bouncing around.)* However, this alone does not cause the Curie Point to change. The dipole linking—which is a quantum effect—doesn't take place in the presence of <u>just</u> a steady external magnetic field when the sample is hot. I'm postulating the internal vector forces created by the changing electric field and changing magnetic field will get those pesky dipoles to re-bond. I'm checking to find out if we can electrically stabilize the dipoles just enough so their mutual linking can resist the thermal agitation that's pulling them apart.

I doubt it'll be a super-effect. The iron can't return all the way to its room temperature permittivity. Also, there is a point where the thermal agitation will become just too much. I'm sure the iron should <u>not</u> get above its Curie Point as there'll need to be some residual dipole linking because those the external fields will only reinforce what is already present.

Remember, just before the Curie Point the exchange coupling bonds are constantly breaking and reforming. Think of this as a way, the four changing fields reinforce a coupling that was about to break apart allowing it to hold together. Plus, some atoms which are almost close enough to reestablish a coupling, get a little extra boost so they re-bond. Quantum mechanics is all about statics, and statically, we intend to increase the number of atoms that have exchange coupling at the expense of those that have broken apart. *(It'll be like an external magnetic field causing iron domains favorable to its orientation to expand at the expense of the others.)*

Also, there is another constraint, one that's given me trouble and is possibly the reason why this effect has never been noticed before. It might also explain why my work with nickel wasn't productive. It could be the effect only becomes significant with metals possessing a reasonably high permittivity. I now believe <u>flux density levels and input energy</u> play an important part.

Those of you who live in cold climates may have seen a long fluorescent light bulb operating in freezing weather. If the bulb gets cold enough, instead of completely glowing, the light reduces to a pencil-thin streak that wiggles and twists along surface of the otherwise dark tube. I believe that when both the current and the flux density are low, those pesky magnetic lines of force can weave their way through the metal core, the small bends necessary representing a slightly lower energy state than interacting with the dipole's magnetic orientation. It's only when the flux density is high and there are many lines of force sharing the same pathway that those tiny bends force each individual line to become straighter and pack closer to one another.

Remember, lines of force repel each other, and concentrated lines of force inside an iron bar will prefer to be straighter so they can fit more closely.

So: when the magnetic flux's lowest energy state is not to weave a path around a ferromagnetic dipole it will keep itself as short as possible and force its way through. And in doing so it will force the electron's dipole into alignment—the same alignment ferromagnetic dipoles have at a lower temperature. *(That's a classical explanation of force energies.)*

Here's a graph showing what I mean. It's a guess, but I think it might explain why no one noticed this effect before.

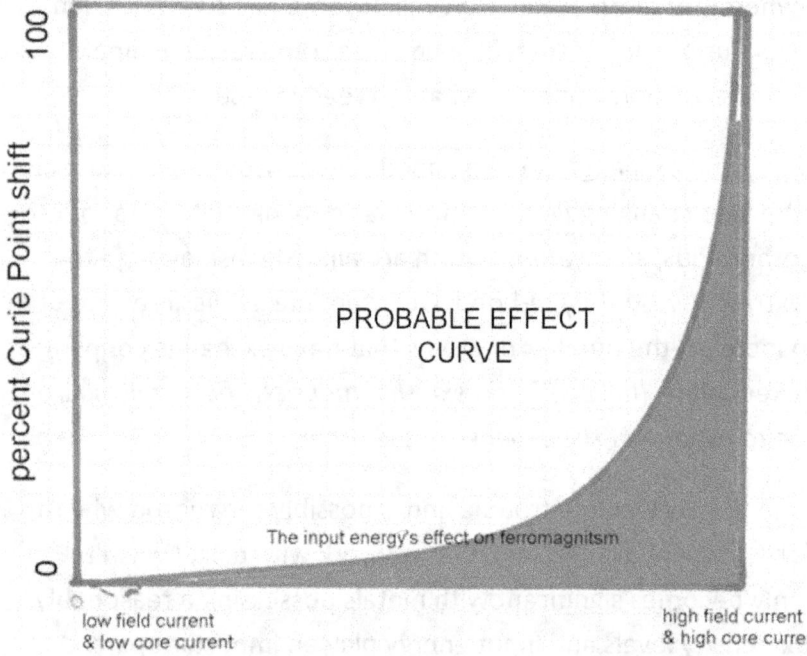

When the electric current through the iron bar and the magnetic flux created by the electromagnet are relatively low, the alignment of those dipoles is barely affected. It's only when the input energy is high the dipoles are forced into compliance. For us, that means the electric current and the magnetic field need to be large enough to trigger the rise in permittivity. You need a strong magnetic field and you have to send a significant current through the iron bar.

CHAPTER 4

THE WORK

Making prototypes to prove the principle.

This is my first reactor core. It's hidden under all the insulation. The magnet coil shown is one of three. Flux is transferred to the core with long, water cooled pole pieces. The tubes carry cooling water once the core gets hot

I've built several prototypes, starting simply and becoming more complex as I discovered various problems. The one above took several years to build, but contained several fatal flaws, which I only understood after it was completed.

The biggest problem I fight with *(as a home shop tinkerer)* is the high Curie Point of steel: 760°C. Nickel's Curie Point is much lower, but nickel is not especially magnetic which makes measurements difficult. Also, I now believe that its low permittivity doesn't allow a high-enough flux density to stabilize the dipoles.

Using mild steel as the heated ferromagnetic insert presents problems. Besides needing to reach cherry-red heat, it is also the <u>same material</u> I'm using to make the field magnets. It's difficult to get the insert up to temperature and keep the field magnets cool at the same time. I suppose that has been my biggest problem. In order for the ferromagnetic generator to have any real-world applications, metallurgy must find a high permittivity, low Curie Point material that can be used as the insert. There are many different alloys that display ferromagnetism, and all have different Curie Points. No one has yet looked for a magnetic material with a high permittivity and a low Curie Point, but it makes sense that with all the possible magnetic alloys available, research could find one which would work.

I'm restricted to using mild steel in my prototypes and trying to find a way to solve the hot place / cool place problem. A second difficulty with using steel is that its permittivity is <u>less than 2000</u>. Transformer iron is around 7000 and most quality magnetic alloys are many times that. The lines of magnetic flux always seek to be as short as possible, and they do travel through the air. I didn't understand how pronounced this problem would become. The above prototype had very long pole pieces made from mild steel. I also wound my secondary pick-up windings around the

primary coil. The result was that little flux made it down the long steel pathway to the hot insert, yet the secondary windings intercepted all the flux that went through the air—overwhelming my monitors with transformer-coupled energy. The above picture, taken while the prototype was under construction, shows the length of the pole pieces. The central chimney, made of series-connected iron spirals, comprises the heated insert, and several inches of insulation protects those field coils. Of course, with those flux-changing spirals needing to reach ~ 750°C and the insulation on the field coils not capable of handling more than 120°C, the only way I had of keeping the coils from overheating was to extend those pole pieces. This problem, and my inability to completely solve it, has doomed all of my prototypes. As I said before, the make-or-break condition for building a successful ferromagnetic generator is finding an alloy with a high permittivity and a low Curie Point so those long pole pieces can be eliminated.

The first picture shows core with its the heating coils. There's an outer heater and an inner one, as the metal needs to get close to 760° C

Here's a side view of the unit after construction. It ended up needing a complex system of cooling. This is a picture of one of the first test runs. The white box under the table is the cooling water drain. Such a simple input arrangement didn't work so well, so the support electronics needed to get a lot more complicated.

Phasing is another issue. The quantum linking between dipoles is reinforced only in the presence of a changing magnetic and electric field. Fortunately, a sine wave conveniently provides two areas of rapid energy change and two areas of minimal energy change. But, for the generator to work, the phasing of the sine wave created external field and the sine wave sending electric current through the ferromagnetic sample must be exactly in phase with each other. I found this hard to achieve.

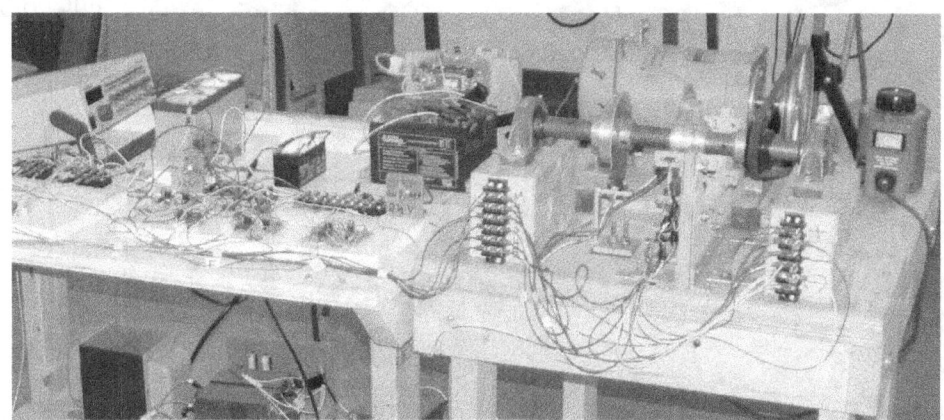

After discovering I couldn't get enough flux through the core at 60 Hz, I tried using this rising-voltage DC saw-toothed wave, but it turned out to be a colossal waste of time and resources. That's the camshaft timer I built to trigger the SCRs I'm used to create a rising voltage source.

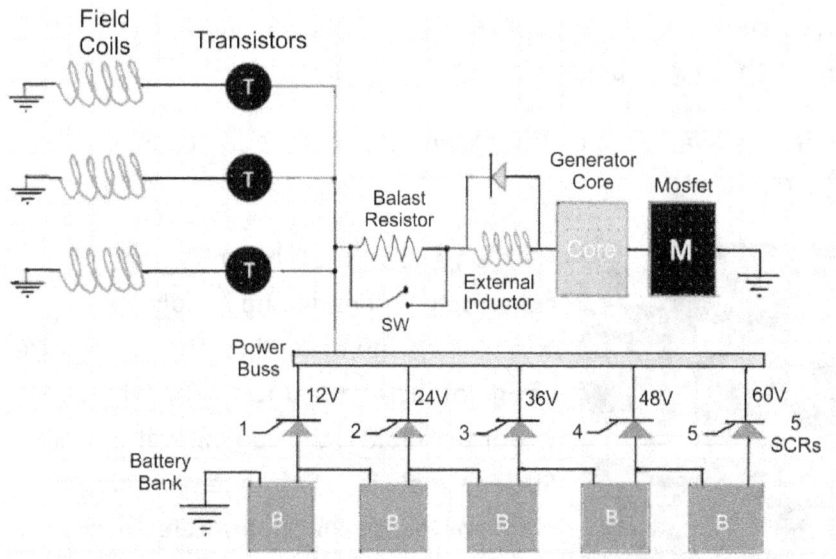

This is the basic schematic of my rising voltage source prototype. It was fun to build, but it didn't work like I expected—mostly because the pole pieces remained too long.

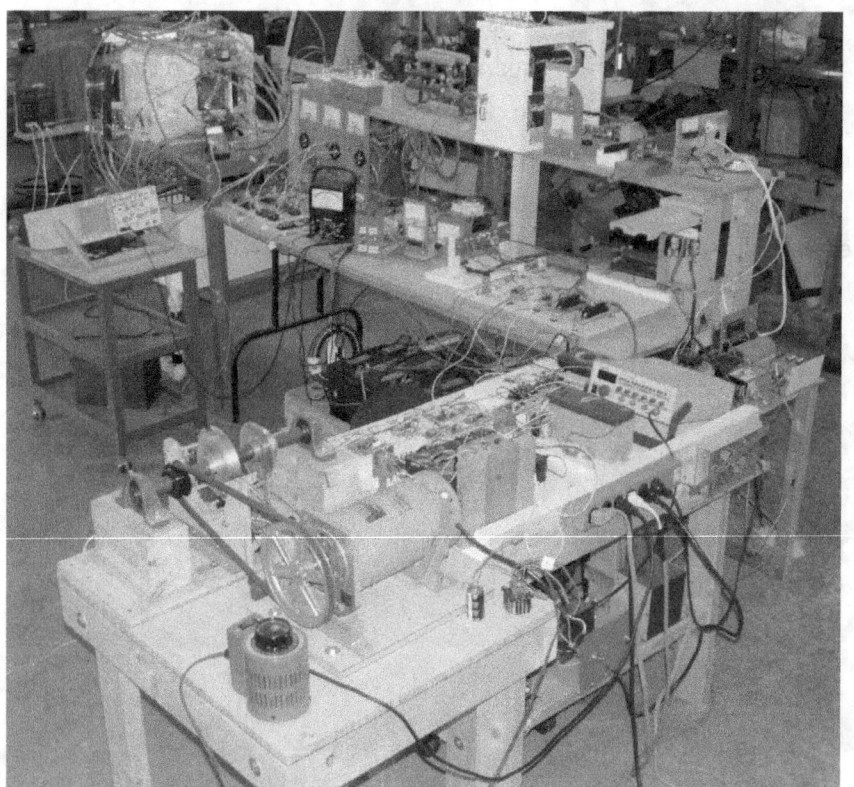

And once everything was completed it looked like this.

It was a fancy rig, but, as I was still unable to get sufficient flux through the heated coils, I finally abandoned the idea.

CHAPTER 5

REGROUPING

After all that work, it was hard to admit that I was never going to get anywhere with those over-long pole pieces.

The only way forward was to start over with smaller models. But, as a tinkerer with few resources, I'm limited to using only mild steel. Meaning the core had to be at ~750°C and in (almost) contact with mild steel pole-pieces whose temperature needed to be kept far cooler. Added to problem of keeping the field coil windings cool enough so their insulation doesn't char, it gets complicated.

I needed a smaller model and went back to the basic idea of an insert between the two poles of a magnet. The ends of the poles could be cooled with water which would allow for only the most minimal insulation between them and the hot insert. But how to get the inset up to temperature? My solution was to use copper strips to channel the heat into the iron sample. Copper is a good conductor of heat, and if one end is heated, it isn't long before all of it is hot.

This next picture shows the idea in its most basic form. When the ends of the copper bars are heated, the heat flows toward the middle to the iron inserts and brings them up to temperature.

Of course, it can't be this simple. The iron inserts need to have an electric current flowing through them at right angles to the magnetic field coming from the pole pieces. And one copper bar will never do the heating job. I need to make the reactor core (insert) of thin iron segments and then stack them together with multiple copper bars pressed between them.

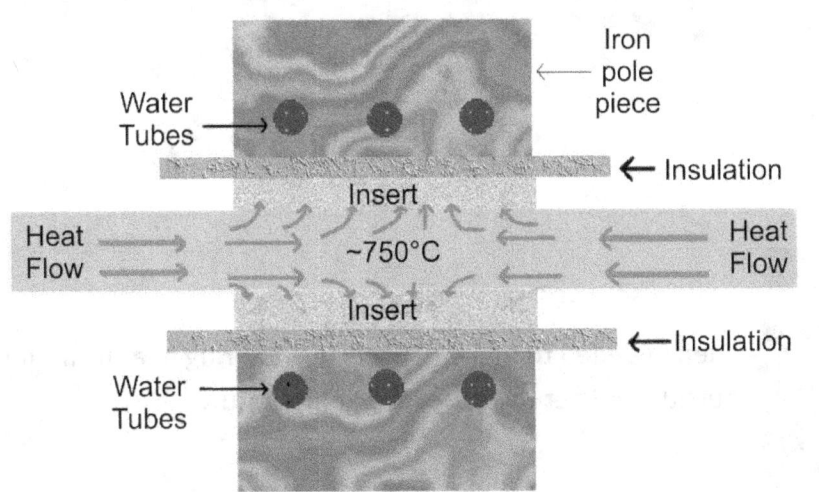

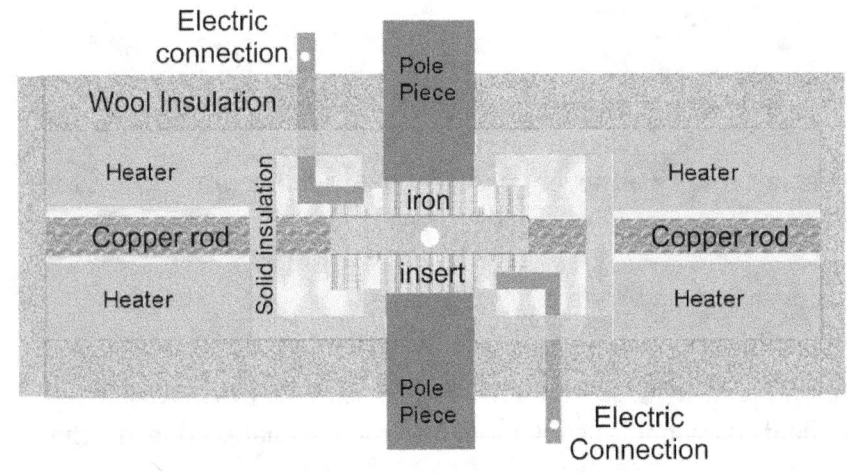

To make the idea work in the real world, I came up with this design:

The iron core is made from 14 segments cut from a 3/32nds inch thick mild steel bar. Each segment is 3 inches wide, by 1 ½ inches high and shaped so current can flow from one side to the other, plus they can be stacked together with insulation so they don't short between one another or the copper rods that are placed between each segment.

To make the electrical connection, I welded long tabs to them so the wire connectors extend out of the hot zone. Each of the 14 segments ends up looking like this:

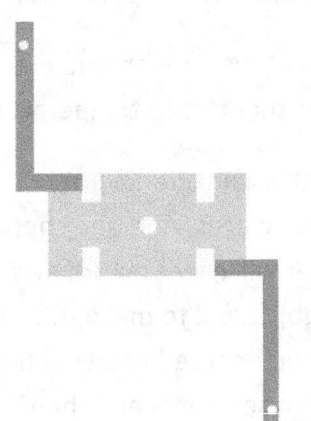

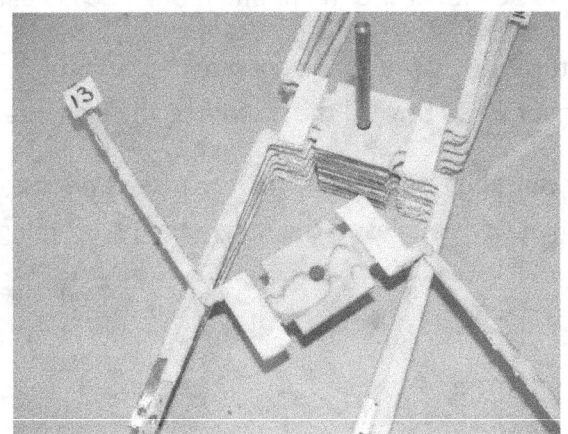

Then I needed to make copper rods to bring heat from the outside and get the iron up to temperature.

This is a close-up of the top of the core. You can see the gaps in the iron core metal where the copper rods pass between each one. Hopefully I've applied enough insulation so nothing shorts out.

This picture shows how the inserts and the copper strips fit together. The unfinished pole pieces are sitting on the top of the iron strips.

The actual iron face-side the magnetic flux passes through remains at 1.5 inches by 1.3 inches.

It looks a bit ragged, but this is how it ended up.

Heat, created by the heaters, should flow into the core through those copper bars. The copper then heats up the core iron. Heat, inside the core, dams up when it encounters the thin insulation between it and the iron pole pieces. To stop the pole pieces from getting too hot, I tried several ways of attaching cooling mechanisms. In the end, I settled on copper plates attached to cooling water tubes. Before buying material and working on the magnetic parts, I needed to see if the core could get hot enough using such a makeshift arrangement.

Building the heaters

Building these two, side-heaters proved to be one of the slowest *(and frustrating)* projects so far. The four heaters are store bought. Unfortunately, I misread the specs on the company's website. The ones I received are not rated for 2200° F but for only 1800° F. While that's hotter than the ~1400F the core needs, I'm sending the heat down copper strips with all the losses that implies. I decided to find out if it could get to temperature before doing much with the other parts.

Four, 300 Watt heaters gives me 1200 watts to send down those copper strips. I would have liked a little more power, but these 300 watt ones are the only six inch long heaters available. The next size up is twelve inches long and that would make the holders huge.

A diagram of what the heaters and the insulation around them should look like:

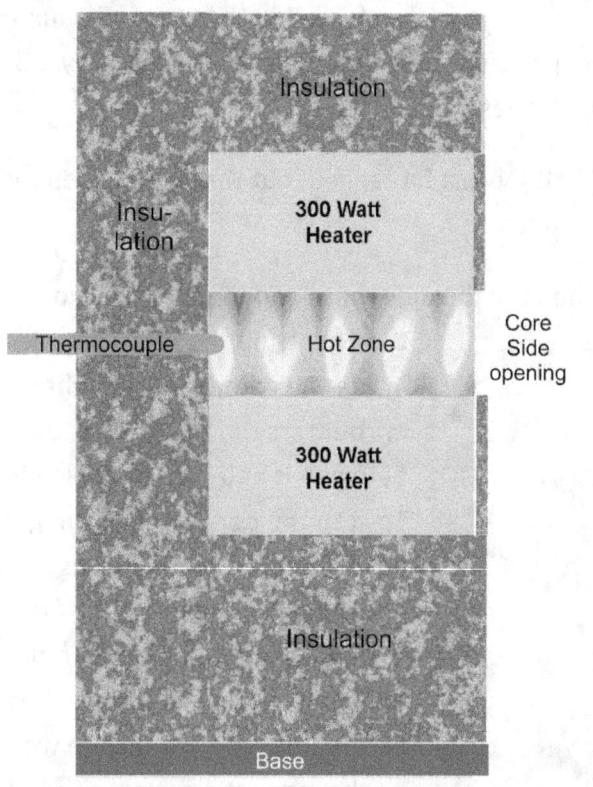

There are two of these, one will fit on either side of the core. The copper rods protrude into the heater's hot zone where they are heated almost to 1800° F.

(Close to copper's melting point of 1981° F—but I think it will be all right.)

This picture shows one of the heater assemblies before putting any insulation around it. There are two heaters, one on the top and one on the bottom.

The hollow core is what the copper rods fit into.

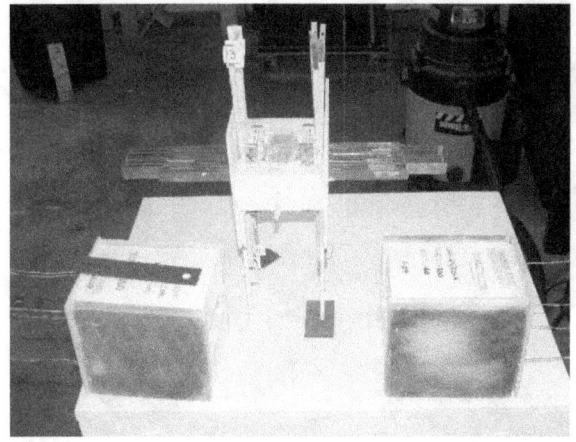

This is a picture of both heaters and how they will fit around the copper rods. I don't want them touching the heater coils, so they will go into boxes which hold them at the same height as the rods.

Here is the heater built into a ceramic holder that keeps it at the height of the copper rods coming from the core.

And the whole thing fits into a box so I can maneuver it and maintain the correct height.

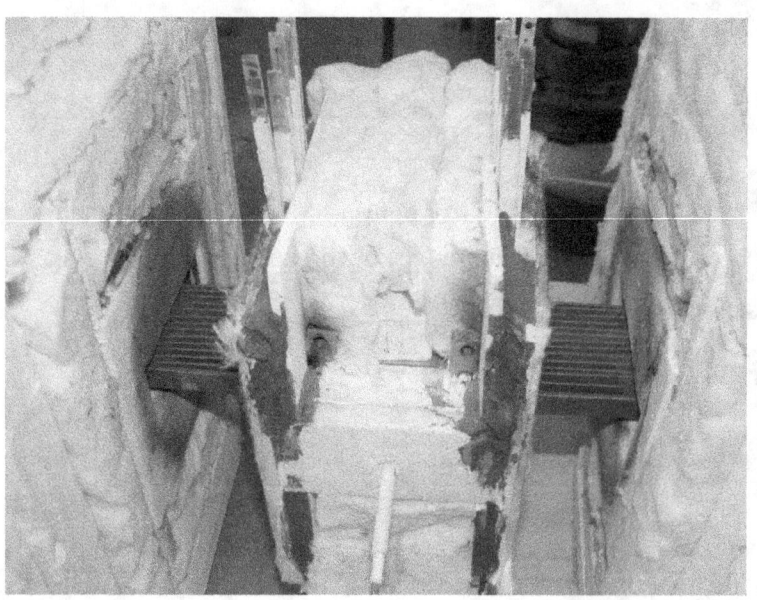

This picture shows both heaters slightly drawn back from the core after the first heating session. I was pleased that they could get the core up to temperature. Here, you can see how the copper rods fit into the heater slots.

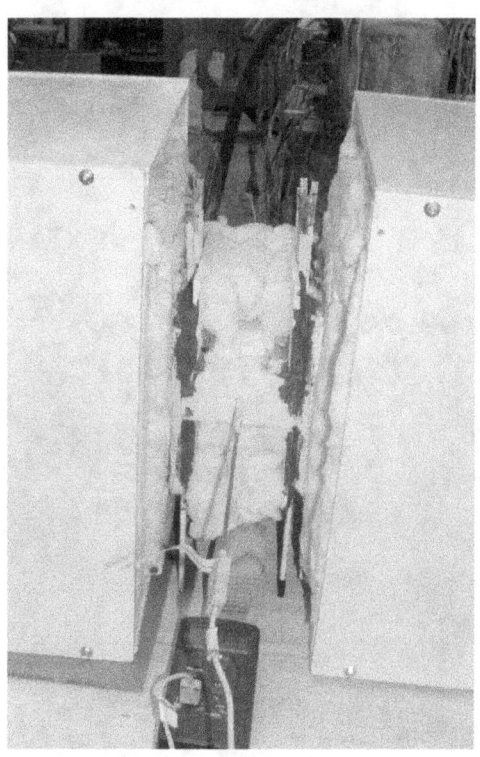

When in operation, the two heater boxes press hard against the core. And it looks like this.

Once I was sure this method could get the iron inserts hot enough, I started building the field magnet and the pole pieces to channel the magnetic flux through the core insert.

CHAPTER 6

The Pole Pieces.

Cooling is a major problem—there is just not enough room in a small set-up to make the copper-tube connections I'd planned on. One reason I built the first prototype so large was so I could cool the parts that needed cooling. For this prototype, I decided on copper cooling fins on the pole pieces which could be connected to a cold water supply.

I made dummy magnet bars out of wood to give me a chance to see how it will go together. Once again, the magnetic flux path appears too long to get much flux through the core. The core will fit between those two cooled pole pieces and the magnet windings will go around the long bar in the back.

 The gap between the two pole pieces should be 1 and 5/8 ths so the core can fit in.

Here, I show one of the pole pieces, which channel the magnetic flux into the core. The interspaced copper fins keep its temperature down as, with only a little thin insulation, it presses against the hot core.

After I finished building two of these, I came to believe the magnetic path remained too long when using mild steel. A magnetic field does travel through the air. All the steel does is concentrate most of it so it kinda goes where you want it. However, all the gaps necessary for cooling greatly reduce the core flux. Made from low-permittivity steel it might suffer from the same problems I encountered with my first prototype.

I had an old 'buzz-box' 115 Volt welder and took it apart. Inside I found a simple transformer. I'm assuming it (might) use transformer iron, instead of mild steel. Remember the permittivity of transformer iron is around 7000 while mild steel is only ~2000. I decided to adapt the transformer and use it as my exciter magnet. An eyeball guess: The input coil has about 170 turns of # 12 wire and the output has about 80 turns of # 8.

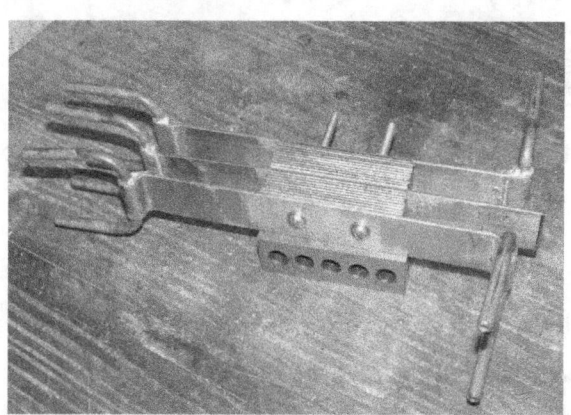

On the downside, using this pre-made frame, the gap between the poles will be 6" instead of the 8" I'd planned on making. I had to discard the original pole pieces as they wouldn't fit and make these smaller ones.

The three copper strips are for keeping the iron pole pieces cool. Note, I made these pole pieces from mild steel and <u>not</u> transformer metal.

This is how I planned to fit them on the core.

The next step is cutting up the transformer: I used a slitting saw to remove the arms on the side that I want to reach out to the core so they would end up being as long as possible. *(Using a slitting saw for something like this is slow process.)*

The buzz box transformer wasn't made with any great precision, so I needed to square the ends. See how that end stack of laminations has a wave in it. To butt it against another piece I'll need to make it straight. This time I remembered to cover and protect the windings while doing metalwork.

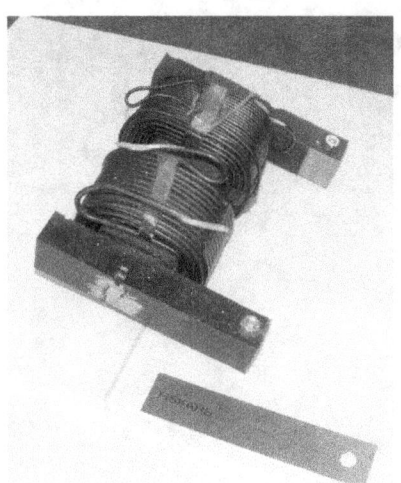

Cutting off the other leg of the transformer was easier because I could use my band saw. *(I didn't need the length.)*

The metal of side I cut off can be used to extend its reach. *(I think its transformer metal, but now my simple tests suggest it might not be. This transformer is from a cheap, light duty, buzz box. Everyone cuts corners when they can.)*

I used the cut-off pieces to make the extensions that reach out to cover the pole pieces. The metal got pretty banged up when I trimmed it, so the laminations are re-insulated with high-temperature white paint.

As my cut-up transformer needed extending, I made two longer bits out of the extra pieces and another cut-up transformer. It took a bit of doing to get everything to fit together even half-way tightly.

Next, I show my new end-bits. Note how they're not identical. One fits on the bottom of the coil and the other is made for the top. Because I can't make the core gap precisely, I've made the top extender adjustable. It can slide down to make tight contact with the cooling pole piece. On the left is the bottom piece. The top one is on the Right. I added some extra transformer iron cut from a relatively large discarded transformer I found.

Here's the bottom piece being attached. You can see how it's made from three sections of laminated transformer iron and how it mates with the main transformer:

I switched things around to get the pieces to mate better. The pole piece with the dark surface has to be on the bottom as it is fixed and stable.

In this picture, you can see the slots on the top pole piece, which let it move up and down.

The transformer partly re-assembled. You can see how I've added two smaller windings. Exactly fifty turns on each leg. I'll use these later to monitor the output signal and to check on the equivalency of the magnetic flux that travels the entire iron-path all the way through the core. I'm hoping this time, as they're not wound on top of the primary coil, their output will more accurately follow the flux that is being influenced by the core.

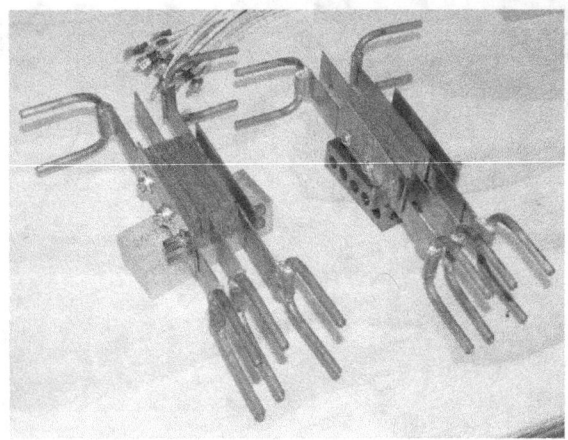

I finished making the two small cooling pole pieces. These are the magnet ends which fit up against the hot core with only a little insulation to protect them. These are not made from transformer iron, just mild steel.

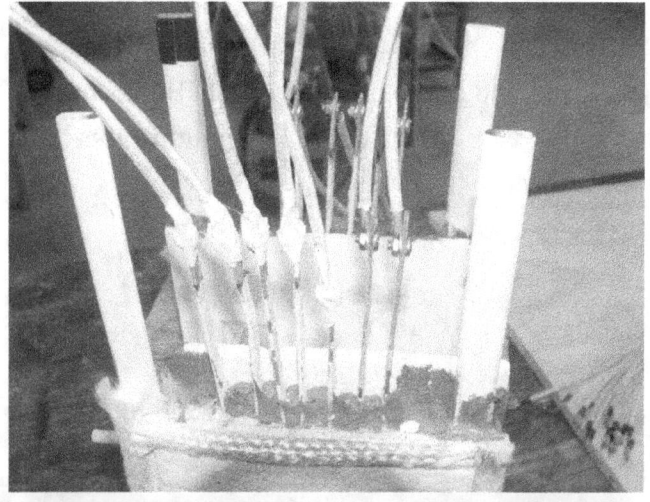

Next, I started to connect the core-current wires to the laminations. Each lamination has an in-side electrical tab and an out-side tab which connects to the tab on the in-side of the next lamination in the stack. As the tabs will get really hot, I'm using high-temperature wire to connect with them. *(It's the same stuff used in your oven.)* Each of the thirteen laminations will require a large loop of wire. This picture shows the underside of the core as the first step is passing the wires through holes in the support table.

Each wire needs to be bolted to the metal tab and then protected with fiberglass insulation. At some point, I'm going to have to carefully label each wire, because this is the kind of thing I'm good at messing up. I wanted to see how it was all going to fit together, so I did a position assembly to see how it would look. You can see how the lamination wires feed down through a hole in the assembly table.

The top cooled pole piece fits in place well. I didn't install the bottom one, but the space for it is there. It's going to be a tricky to get that bottom pole piece in place because it has to be "pushed up" against the bottom of the core.

(Ceramic insulation is not a strong material, and I don't want anything to break.)

I'm coming to the end of the building phase. Here's how it looks after I bolted on the top set of lamination wires to the core assembly.

The Table

The two slits are to let the bottom lamination wires have someplace to go and not get squished.

I'm using wood this time. Now I'm more comfortable with the power of ceramic wool insulation, I find wood is acceptable. Plus, it's way cheaper.

The first item on the table is the riser. The underside of the heavy field magnet needs to be pushed-up against the bottom of the core.

These are two pieces of wood cut like wedges, so their combined height is slightly adjustable. The little strip of aluminum goes between the two wedges allowing them to slide across each other easily.

On top of the riser I place the half-assembled magnet. The pole piece you can see will press against the cooled pole piece attached to the underside of the core.

After that, I bring the wired core, with its underside cooled pole piece attached, over and line it up with the magnet's pole piece. In this next view, you can see how the bottom wires feed through the holes in the table.

The next step is installing the upper cooled pole piece. Remember the pole pieces need to be bolted to the core to prevent the motor effect from ripping the whole assembly apart. Those

bolts and tabs are way down in there, so it's tricky.

In this picture, you can see the riser screw. Tightening it pushes on the wedge and this raises the whole magnet assembly.

The upper pole piece attaches to the magnet and by its weight presses down on the upper cooled pole piece. This is why I made the attachment for the upper piece differently from the lower one. The upper one can ride up and down relative to the magnet.

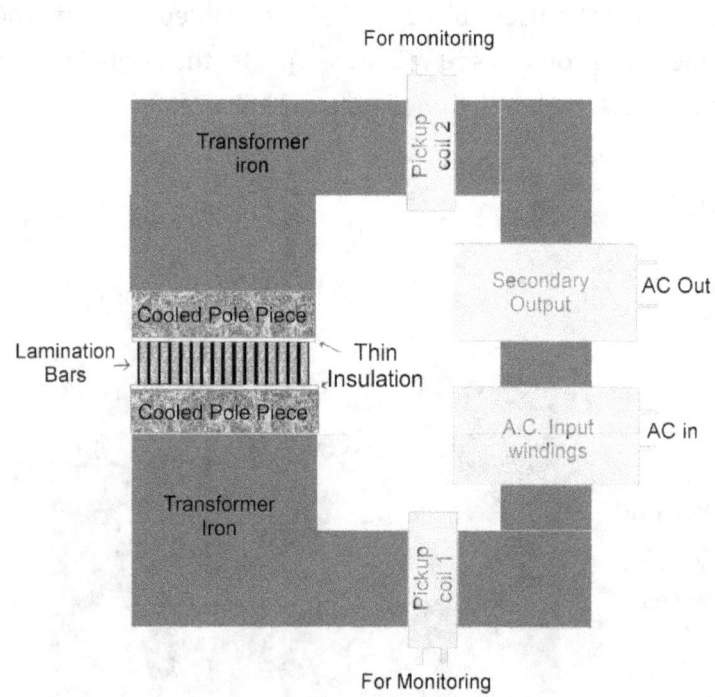

O.K. It seems to me these pictures appear more complicated than necessary. This is my simplified diagram of the core and coil looking from the side. Reduced to the essentials, this is a simple transformer with a small insert inside its magnetic path. I'm also using the secondary transformer windings to create a current for the lamination bars of the insert.

The secondary is only a few turns of a heavy wire, so it can produce a lot of current at a lower voltage. I installed two smaller windings to monitor what is happening when the insert gets hot.

CHAPTER 7

Electric system.

Once I had the magnet on the table, I started checking the electric system. When it was a welder, the transformer input was 120 Volts. To begin with, I'll use 60 Volts and work up if I need to.

I should have measured the no-load current on the input coil before cutting the transformer apart. Now, in its new arrangement, I find the input side draws five amps at 60 volts when the output coil is <u>open</u>. That's <u>a lot</u> more than I expected. It would have been nice to know how much my cutting it apart and reassembling it with an insert degraded its transformer properties. *(Most transformers draw next to no current when they're not passing energy to the secondary. However, while I'd expected a cheap welding transformer with a duty cycle of around 20% to draw a significant idle current, I wasn't expecting it to be so high.)*

I put some resistive loads onto the secondary to duplicate the slight resistance of the laminations after I get them wired in. Two of my 1 ohm resistors connected in series gives me 2.2 ohms. When they're across the secondary, at 60 volts, the input side draws 6.5 amps and the output shoots 8.5 amps through the resistors.

Just one of the 1-ohm resistors draws 14 amps from the secondary. And the input coil is then drawing 8.5 amps.

On the bright side, it looks as if, even at 60 volts, I can get a significant current to send through the laminations. Unfortunately, once again, I'll be dealing with a large amount of input energy that will swamp any small change due to the ferromagnetic effect.

Some more boring numbers: with the core connected and wired up so the laminations feed one to the next. At 60 Volts feed voltage, this is what my preliminary tests show: *(I'm stuck using 60 Hz house frequency)*

The input coil lets 14 Amps flow through it. The output coil then sends 35 amps at 6 volts to the stacked core laminations. That means, at room temperature the core laminations all wired together have a total resistance (reactance) of 0.017 ohms. Although at 730°+C I expect the core resistance to rise. (Dropping the current.) I'll work with 60 volts to start.

(With these high currents, if I don't use ballast resistors on the input line, I'm a bit limited on

possible voltages. I own two 120V to 60V step-down transformers, and wired in parallel they can give me the 14 amps the input coil needs. But my autotransformer, while it can produce any voltage between 1 and 140, has a top limit of 10 amps. So, unless I invest in some expensive equipment, I'm limited to 60 and 120 volts. But at 120 volts, the input would draw 28+ amps from my wall socket. I want to avoid that if I can.)

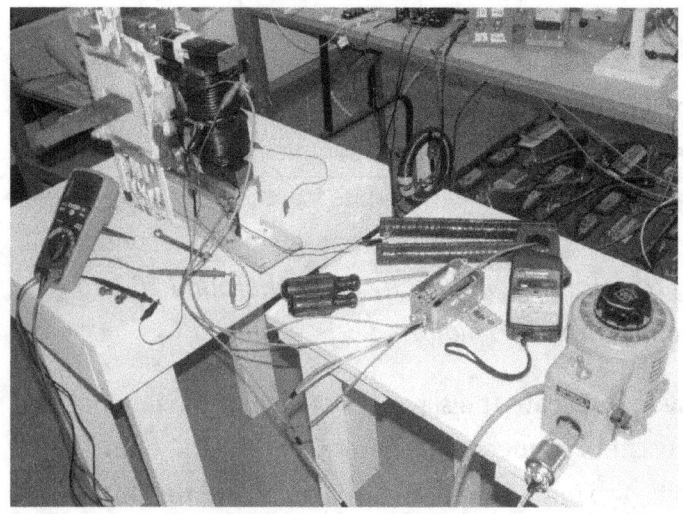

This is testing the amperages and outputs at various voltage combinations.

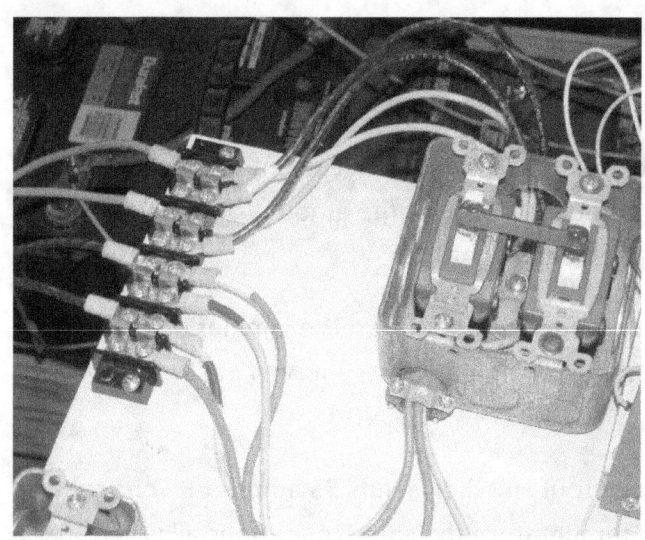

A picture of my simple core-current phase reverser. The AC current from the output coil comes in on the two wires at the top left. It then goes to the switch box and then back out the black wires leading to the core. One of the switches in the box is mounted upside down and they're both ganged together, so if one is on, the other must be off.

* * *

Now it's time to install the two heaters on either side of the core.

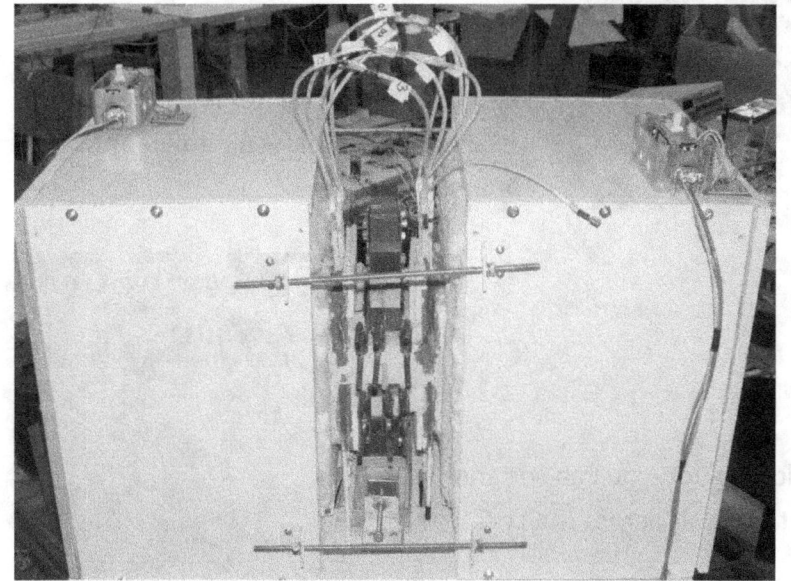

Because the heater shrouds are wood, they hide the working parts. In my first tests, I just pressed them up against the core, but to get a better seal, I've installed threaded rods to pull them together and squish the core between them.

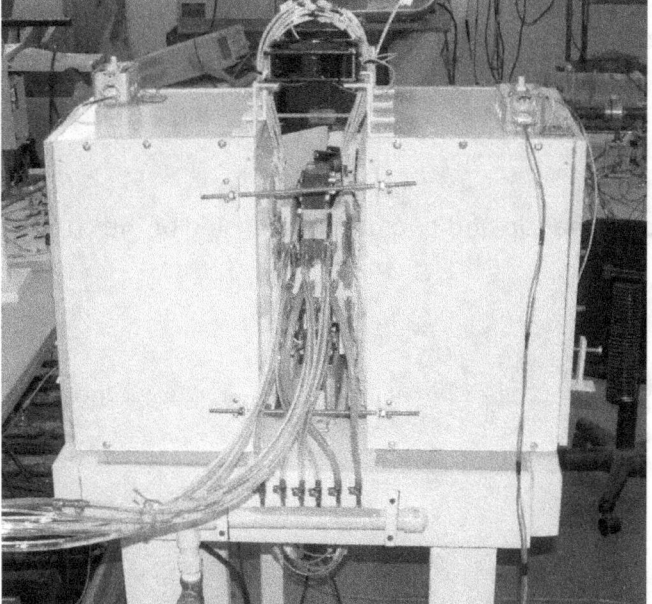

On top of the boxes are the switches which turn the heaters on and off. Remember, those two heater sets are not perfectly matched. I'll have to monitor each heater temperature and manually turn them on or off as needed.

(I put LEDs by each switch so I'd know if the heater is on or off.) It sure is getting hard to see anything.

Now it's time get the water lines hooked up and install the fans that will blow air over the coils in the rear. The water system went in easier this time. It only had six lines. Unfortunately, this only cools the front end of those copper bars I put in the pole pieces. I'm using the same drain system as I'd built before, so I only have to extend the waste lines over to it. But, since the water-cooling is minimal, I've decided to increase the fan cooling. I'll install four fans, two on the top and two along the back edge. This means the core and the field magnet are going to be pretty much hidden once it's all put together.

This is the top fans after I got them installed. They'll blow air down onto the core.

Looking from the backside of the field magnet, you can see the two other fans (the black things) that blow air across the core.

There are four fans total and that is in addition to the water cooled pole pieces.

Here is a closer look at the fans. There are two on the top and two on the rear side of the core-magnet assembly.

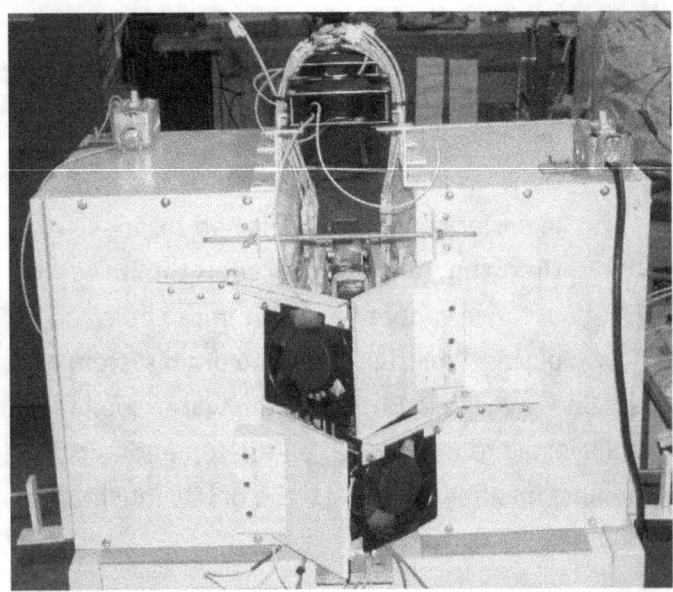

In this picture, you can just make out the field magnet and the core squished between everything. Note the threaded screw which keeps the two heater boxes pressed against the core.

Here is a side view taken after I removed one of the heaters. It shows the set-up a little better as the fans have been removed.

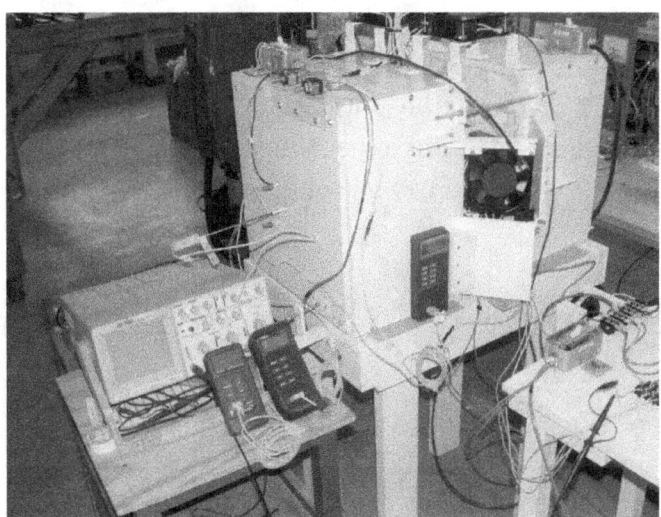

A picture of the core squished between the two big heater boxes with four fans ready to keep the metal cool.

Three thermocouples keep track of the heater temperatures and the core temperature. All the output monitoring equipment sits on the tables next to the core.

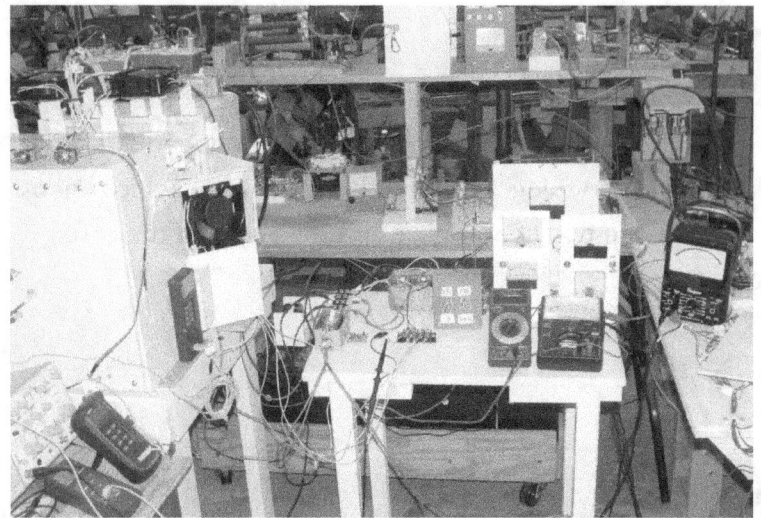

First job: measure the input and output volts and current with the core at room temperature. I have two small output monitoring coils wound around the transformer iron and I'm using them differently. One takes the AC from the coil and puts it across a 20-ohm resistor. Besides measuring the volts and amps, I can view its output on the oscilloscope.

The AC current from second monitoring coil I'm sending through a diode bridge and turning it into DC. That way I can observe its output as watts and see if power coming from the coil increases or decreases. This is to help me distinguish between any power created besides that generated by the transformer-coupled output.

I'm also measuring the amps produced by the main secondary. This is the current I send through the laminations. When I apply 60 volts AC to the primary coil, it draws 12.5 amps of current. That means at 60 cycles the inductance of the primary acts as if it were a 4.8 ohm resistor. Through magnetic coupling, the current through the primary creates a larger current in the transformer's secondary coil. This is the current I'm sending through the laminations.

I placed that reversing switch I showed before on the secondary. This is so I can send the current through the laminations in two different directions. *(Since its AC, what I'm really doing is changing the phase relationship of the two sine waves.)*

At room temperature, with the reversing switch one way, I'm arbitrarily calling that position left; the secondary produces 7.89 volts and it can send 30 amps through the laminations. When I flip the switch the other way called "Right," the secondary creates 9.27 volts but can only send 29 amps through the laminations.

There is always transformer coupling. That means my two monitoring coils will produce some power whenever current flows through the primary.

This is where it gets depressing. The two monitoring coils each have the same number of turns as the main primary. In a perfect world, they would each put out the same voltage and amperage as the 8-gauge primary which I'm using to power the laminations.

(Using resistors on these two output coils keeps the current down to something more agreeable to the wire's lighter gauge.)

The only difference is the <u>physical position</u> of these coils. The thicker-wire main secondary is on the same transformer leg as the primary, but my monitoring coils are wrapped around the legs which extend toward the core.

Unfortunately, the resulting coupling was really bad. Reversing switch right: monitoring coil: AC output voltage 11.5 RMS volts. 0.6 amps into 20 ohms. That's a measly 6.9 volt-amps.

Rectifying the output to DC lets me observe watts. Into a 10-ohm load, the output coil gives 0.4 amps at 3.7 volts. That's a pitiful 1.48 watts.

The power coming out of the two identical monitoring coils should more or less be the same. Putting the reversing switch "left" gives slightly different, but depressingly small, numbers.

This shows very little of the magnetic field produced by the primary is being diverted down the long legs and though the core. I guess most of the generated magnetic flux finds it easier to cut through the shorter air-path than to follow the transformer iron all the way around the circuit.

Magnetic flux is tricky. Its primary goal is to always remain as short as possible and it does go through the air. The iron provides a preferred route and the flux will extend itself to follow an iron path. But only up to a point. Any small air-gaps in the iron path quickly negate the iron's desirability. *(Alas, my poorly made core insert does represent a substantial air gap.)* Before even bothering with the heaters, I should have spent the time to determine what the magnetic coupling was like and if this was even a reasonable configuration.

I've redrawn that picture of the basic coil arrangement, this time including those tabs I couldn't cut off. Then I shaded the area showing my idea of what the magnetic field generated by the primary coil might look like.

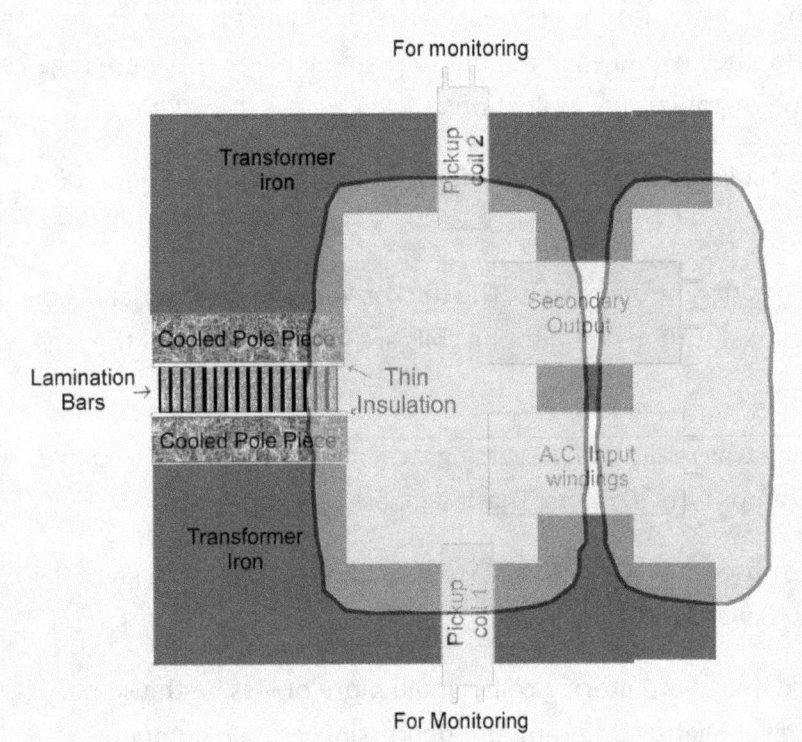

Assume most of the magnetic flux created by the primary coil is concentrated in the shaded areas.

Also, remember this is a two-dimensional representation of what is happening in three dimensions. Essentially, most the magnetic flux has little reason to follow the iron path though the core.

As the cooled pole pieces and the laminations (made from regular steel) along with the insulation act to reduce the flux, it makes sense much of my generated field is taking the shorter route and mostly forcing itself through the air surrounding the primary coil and so avoiding the core.

This prototype has a much shorter magnetic path than the first one I made. Plus, the first one was completely built from mild steel with a permittivity of ~2000 instead of transformer iron's ~5000 TO 7000. It's no wonder my first model didn't show any reaction.

The upshot of all of this is—I suspect little of the generated flux is diverted down the legs so it can interact with the current in the laminations. I doubt this prototype will reach the flux density inside the core I believe is necessary to reveal a Curie Point shift. Still, some flux must be getting through the core: Flipping the reverser switch does change the voltage and amperage of the lamination current. It must be like this: what little flux does make it all the way to the laminations from the primary coil is being 'pushed' inwards by the core current.

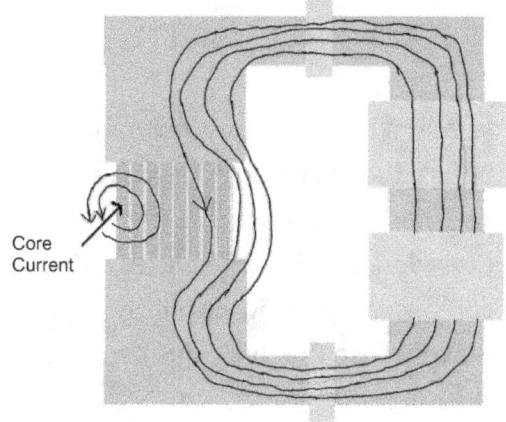

This first simple drawing <u>shows only that few percent of</u> the flux which makes it all the way around my magnetic circuit. The core current is pushing the flux outwards, making for a longer path.

And when the reverser-switch is in the opposite configuration it would look like the next picture.

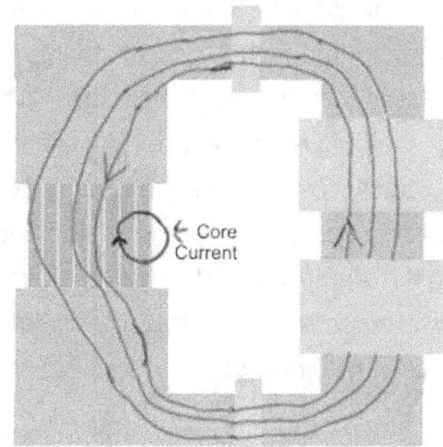

This is the reason I can see a small difference in voltage and current when I flip the reversing switch. It at least tells me that a little flux is making it to the laminations. I'll try a few test runs and see what I get.

OK. I have a notebook full of measurements. There are slight shifts of course. Heating the core to 1400°F is bound to create a few changes. But, in that critical zone where I'm hoping to see something happen with the Curie Point, I can't point to anything decisive. The oscilloscope may show a tiny increase, but it's too minimal to claim success.

One thing surprised me: Heating the core up above its Curie Point—to 1500°F or more, so the insert could no longer be ferromagnetic, <u>didn't</u> show a change on either the input power or the output voltages!

When the core is above its Curie Point, the magnetic path should encounter a 1 and 5/8ths inch air gap that wasn't there at room temperature. We know any air gap decreases the flux in the magnetic circuit. *(That's the main principle behind this generator and it's a well-established rule.)* Even with that big heat-created air gap, the current in my prototype doesn't reveal any change! I can only assume it means <u>next to no flux</u> is following my metal path to complete its circuit. It's all *(within the boundaries of my detection equipment)* using the shorter path through the air.

I've built another prototype with a major flaw.

Here is a picture of the magnetic set up before the cooling pole pieces and the core laminations are inserted. Considering that those pieces and the core insert have low permittivity, it makes sense that little flux completes the long path.

I rushed building this one far too much. I should have spent more time examining what kind of transformer I was creating from that cut-apart cheap welding rig. Ideally, even before bothering with the heating system, I should have made a rotating insert. It would have a steel side and a non-magnetic side and, as it rotated between the pole pieces, then I could have seen if it significantly changed the overall flux. *(Even cleverer would have been to make some measurements on the welding transformer before I cut it apart! That it now draws such a high "idle current" can't be good.)*

I haven't examined everything. Besides not getting much flux through the core, there are many other variables; phasing, current and flux density, along with numerous physical arrangements. I can't even consider doing metallurgy to see if this effect is somehow dependent on the metal's crystal make-up. All of these things should be examined more critically. My main obstacle—that my field magnets and core have the same problematically high Curie Point, could be overcome by someone with access to good metallurgy. Or with sensitive instruments, it might be possible to confirm the theory using a core made from nickel.

Just the idea that it's possible to manipulate the Curie Point, might get more scientists and engineers interested. The crisis of World War Two encouraged scientists, engineers and even whole corporations to put aside short term profits and work for the common good. And many of the ideas they pursued, (a very large percentage) didn't amount to anything that shortened the war. But that generation was willing to take that hit to find those few breakthroughs that could alleviate human suffering. Global warming and worldwide economic collapse has the potential to make even WW2's horrific death toll seem small.

I'll keep plodding on, hoping to discover something to convince others to examine this idea more closely. But the reason I rushed building this model is because my health and money are no longer the best. I'll have to see if I have yet another prototype in me.

Let me reiterate; already there exist many underutilized ways of producing electrical power from sources like solar, wind, tidal, geothermal, and hydro plus a few far-out ideas like mine. The ferromagnetic generator is not a last-ditch stand. And keep in mind oil and coal's main attraction is only its current cheapness—as long as our next generation assumes the expense of us oxidizing so much carbon.

In a world which still allows hedge funds and derivatives because they allow insiders to exploit short-term profits over true investment, it's hard *(or maybe impossible)* to change the thinking of those who make decisions. That's why I feel it's necessary for those few who care about future generations to pursue off-beat ideas like the ferromagnetic generator.

CHAPTER 8

I've decided to make another attempt to find out if it's possible to change the Curie Point electrically. My health has improved and I'm now managing to spend more time in my shop. I've heated up my second prototype several times, and each time I've become more convinced it doesn't work because there's so little MMF (flux) making it through the core. Occasionally, when the core heat hovered around 1400°F, I thought the oscilloscope showed a widening around the top of the sine wave. But it wasn't anything provable. I keep reminding myself, for this method to work, both the core current and the magnetic flux must surpass a certain minimal density. With so little flux passing through the core, I know I'm not meeting that requirement.

At this point, since I'm only trying to prove the idea, I've decided to break <u>many</u> good-design rules. I'm going to eliminate the cooling problem. Instead of using water to cool the coil/transformer and its windings, I'll separate the parts and heat up the core laminations independently. Then, when the core metal approaches ~1400°F, I'll shove everything together. That'll give me a few minutes of working time before the transformer windings begin overheating. Not a very elegant solution, but there's a chance it will be enough to prove the concept.

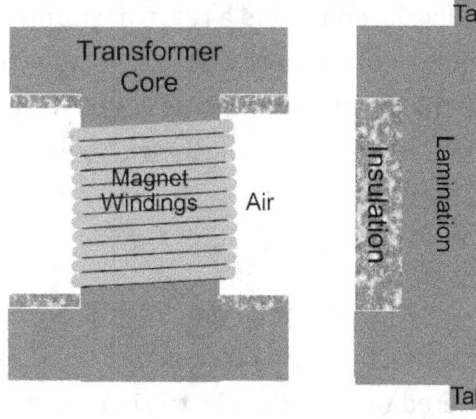

Here's a picture of what I'm thinking of building: two independent parts. The exciter magnet remains at room temperature while I heat the core laminations almost to the iron's Curie temperature.

Of course, it'll need lots of insulation around the heated parts, and I'll have to find a practical way to heat up the laminations. To make it easier, and use what I already own, I'll use that trick of using copper rods to funnel heat into the laminations. The trick worked for me last time so maybe I'll get lucky. I only show one lamination in this side view, but there will be many once I get the dimensions figured out.

There is a <u>big change</u> in this design. I debated long and hard about making it. But as my previous work hasn't proved fruitful, it's a chance I've decided to take. If you look at the design, you'll see I've not drawn the core laminations at a right angle to the expanding magnetic field. This time the electric current will run parallel to the magnetic field.

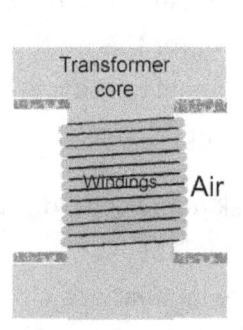
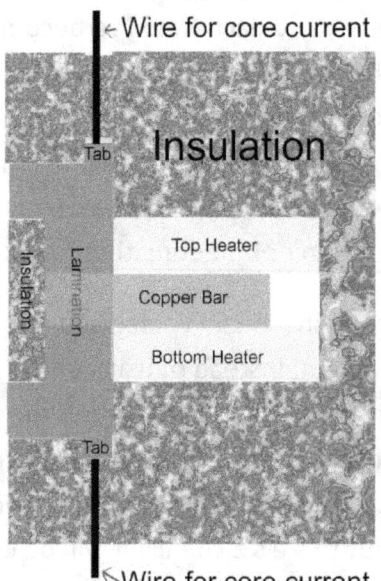

Thinking about the internal electric forces affecting the dipoles, I can see how this arrangement might provide stability for them, and making the prototype this way will be much easier. Understand, I don't expect this orientation to be as effective as crossed fields, but since it's a quantum effect, there should be some interaction.

Here is the way I plan to heat up the core while keeping the coil windings cool.

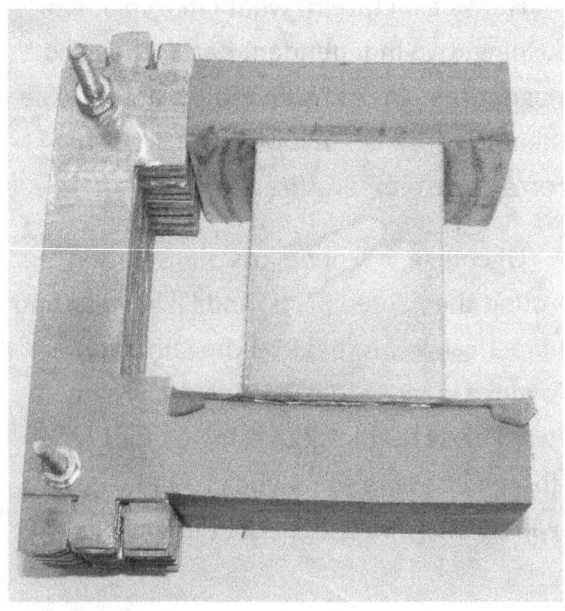

I was luckily and found a real transformer of approximately the right size and cut it up and removed the old copper windings as I only needed the internal metal. Then I machined new side pieces to butt up against the original transformer center.

Yes, the core came from a real transformer, but alas, my laminations are *(once again)* mild steel. I cut staggered tabs for the electric inputs so there'd be room to make the connections without shorting one to the other. Remember, when finished, the laminations will be insulated so the current flows the long way through each one.

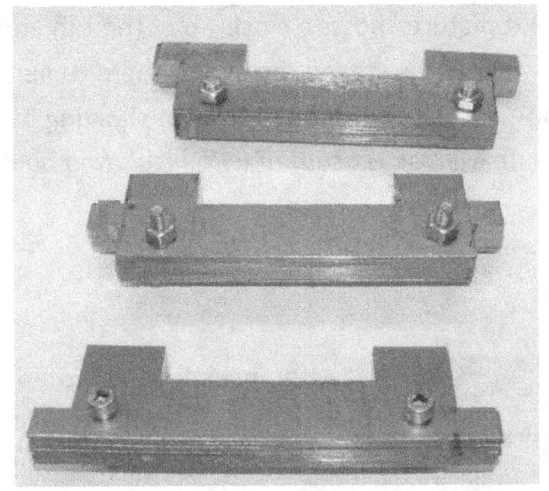

Just to make it easier to understand, here are the laminations after I've machined them. This shows the three different positions for the offset tabs.

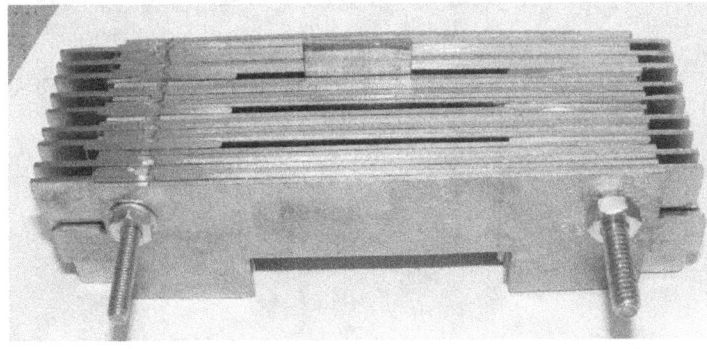

Here, I show a set of assembled laminations viewed from the back where the heat will be applied. The gaps are for strips of 1/16th thick copper bars needed to channel heat into the iron.

(I put a bit of scrap copper into one of the slits so you can see what I mean.)

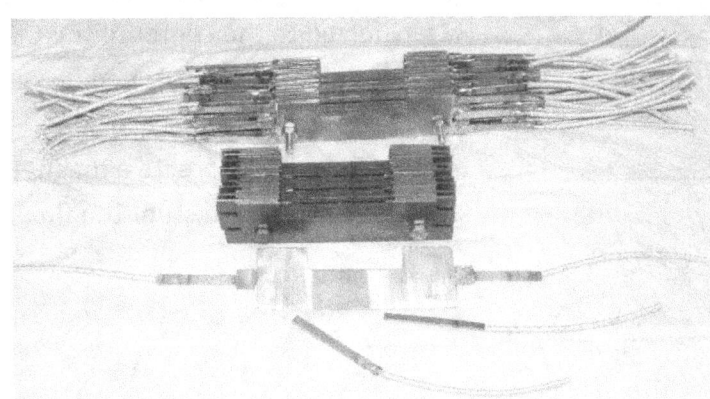

High temperature wire's insulation is only good for around 500°F. I made metal standoffs to extend the lamination tabs, which I hope will keep the wire's temperature down. The following picture shows all three stages so you can see what I mean. I haven't yet painted the laminations with insulation.

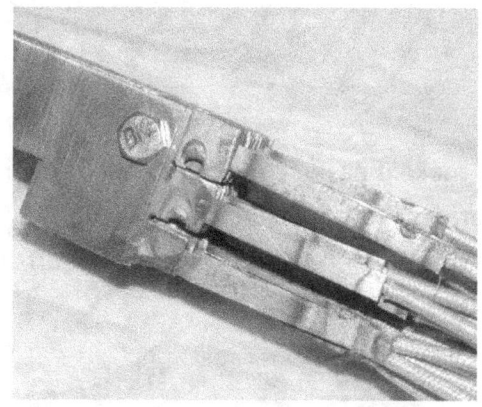

This next picture shows a close-up of the tab ends. It's to reveal how the staggering allows them to lie close to one another but (hopefully) not shorting out one to the next. *(Not yet coated with insulating paint)*

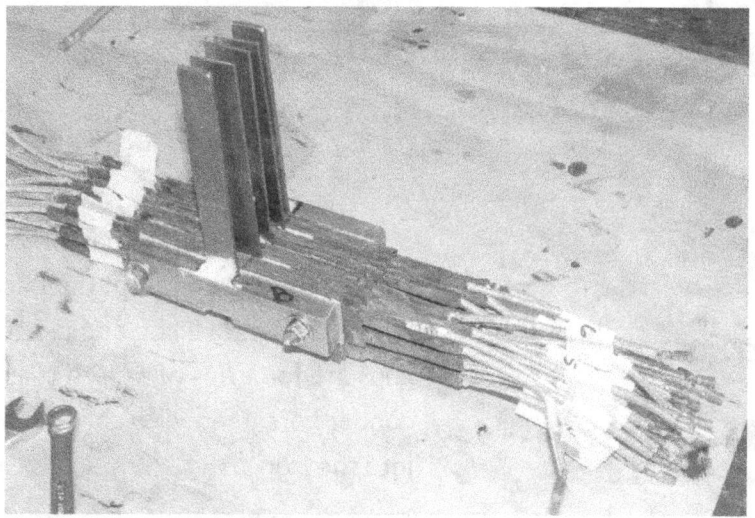

After insulating, I stacked the laminations with the heat-input-bars inserted every half inch. The heavy bar which presses everything together is stainless steel, so it doesn't provide an alternate magnetic path.

And here's a view from the front.

Heat, picked up by the copper bars is transferred to the center of the C-shaped laminations. I only need a small area in the center to reach around 1450°F. That should create a magnetic "air-gap" enough to disrupt the flux.

I've rewound that real transformer core to meet my needs. It has a primary coil and a lighter secondary coil, which I'll use for monitoring, wound on top of it.

To show how it fits against the C-shaped bars, in this next picture, I'm showing it in the horizontal position.

The copper rods sticking out the back of the laminations fit into the heater cavity. Here, I have the heater box lying on its back and I've placed the transformer core on top of the C-laminations so you can see how one side fits together. I'll need to provide a way to hold everything together when the heater box is turned upright.

I made brackets to hold the core against the heater and insulated the front.

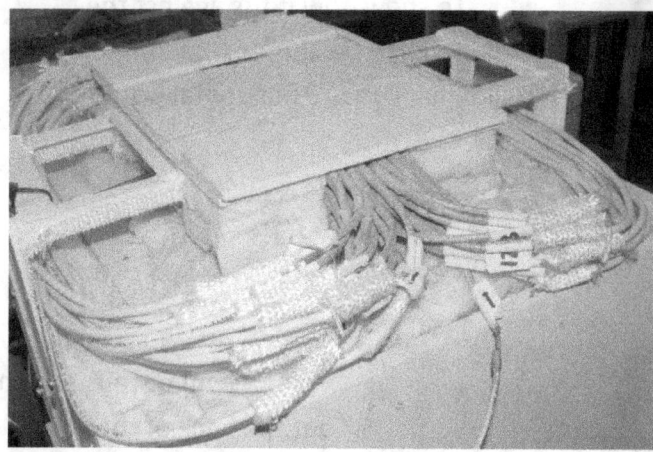

And once all those lamination leads are connected—the bottom of one to the top of the one next to it. That way the current goes in on the top of number one, out the bottom and into the top of number two, and etcetera. This allows the current to go through each lamination in the same direction. Having all the current flowing in the same direction might not be important. But, with so few chances left to experiment it's best to play it safe.

This is one heater finished; I'll need to build another for the other side of the transformer core.

After testing.

Bummer. The 300 watt heaters can't get the core hot enough. It tops out at 1200° F. I suppose the insulation isn't sufficient as the exposed tabs on C-laminations lose so much heat. *(I measured one at 700°F)* And there is only ½ inch of insulation over the main (hottest) part of the core which leaks a lot of heat. To make it work, I'll need to increase the insulation over the core and do what I can to make the rest more insulated.

I only need another 200 to 300°F. It might be possible to get the laminations hot enough if I attach tab extenders to the metal core. Making the tab ends longer will let me increase the central cavity insulation which right now causes a major heat loss.

Trying to increase the core temperature.

To save time and hopefully raise the core to the Curie Point, I've decided to continue using the core assembly I've already built. With extenders, I'll have an inch of insulating wool over the central hottest part, and that, with better insulation around the sides might get me those last few degrees. I'll also make removable covers to put over the exposed ends on the core while it's heating up. With any luck, these tricks will be enough.

I'm not happy about using these extender tabs. Naturally, they don't fit as tightly as I would like and so create yet another air gap. Here I am gluing those tabs onto the core's ½ inch extensions.

With those extenders in place, I can put twice as much insulation over the hottest part of the core. I've also increased the insulation in any other place I could find.

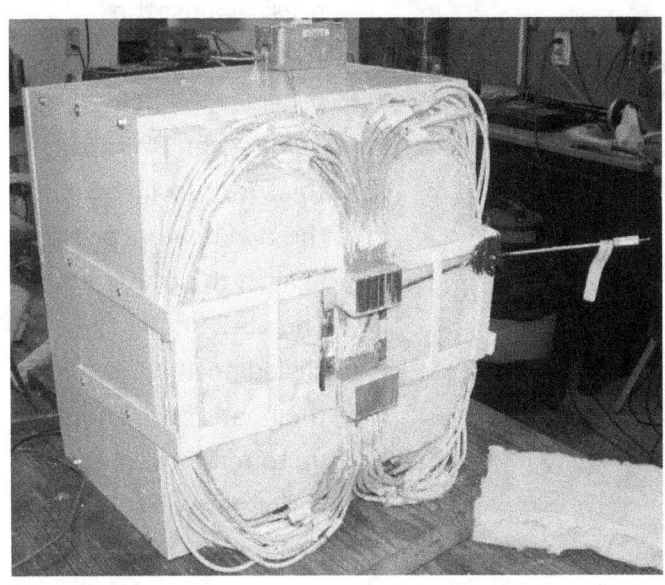

In this picture, you can see how the core metal extends farther out with those extra tabs in place.

After the extra insulation is in place it looks like this before I put insulation in the central hollow cavity.

There are two sides to the transformer, so I need two heater set-ups. I built a way to press them both against the sides of the transformer on a rail, which allows them to be moved back and forth and end up positioned where I want them.

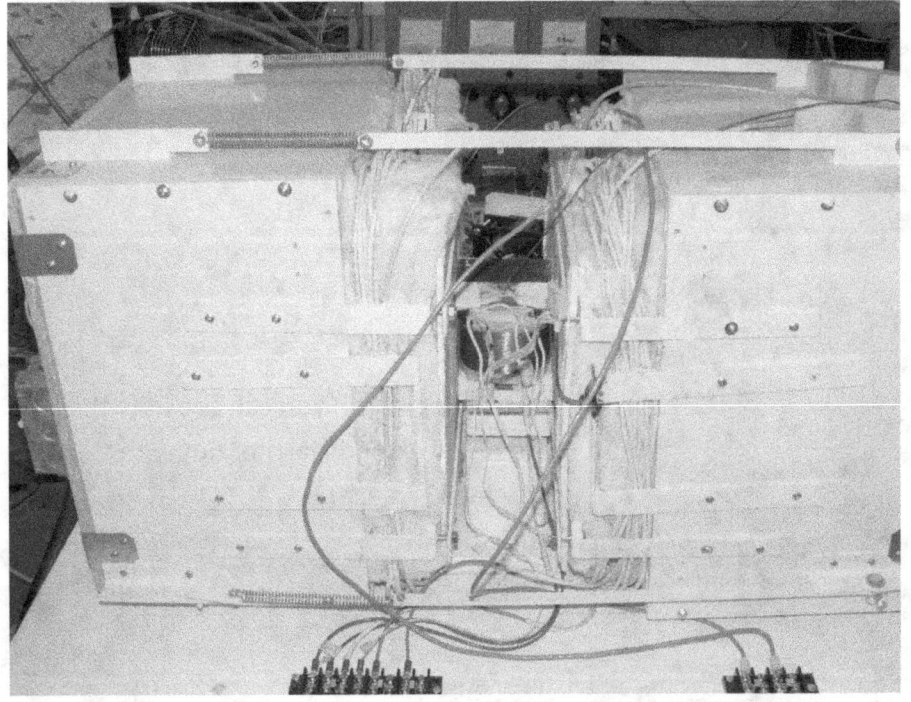

This picture shows how the two heater/core assemblies press against the transformer that provides the field magnetism. On top, are the springs which keep the two heaters pressed against the transformer. This view shows the closed position. While the heaters are coming up to temperature, the springs are disconnected and they're backed off from the transformer and its delicate magnet windings by several inches.

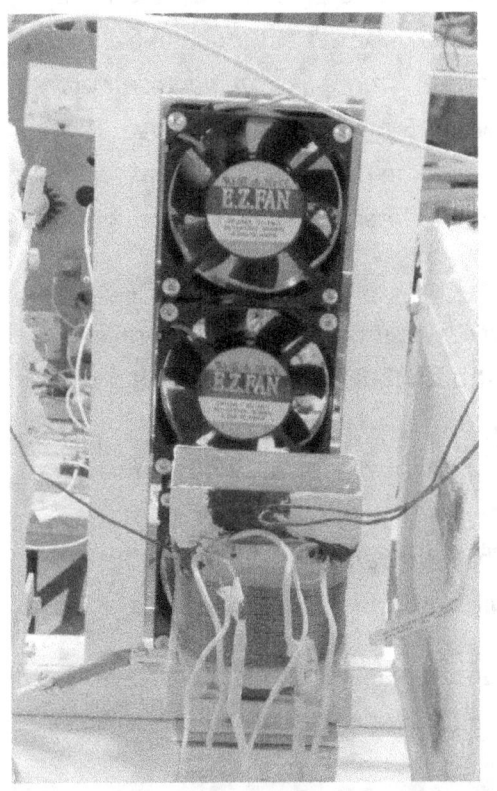

Using a prototype in this fashion will require precise before and after measurements. More on that when I reach that stage. Meanwhile there's a bunch of little details to handle.

Here's the cooling tower: There are three fans and they blow a stream of air between the two heater assemblies to keep the transformer windings cool. While the two cores are coming up to temperature, I'll insulate the core's exposed ends. That will let them reach temperature more quickly and the fan's air-stream won't be working against me.

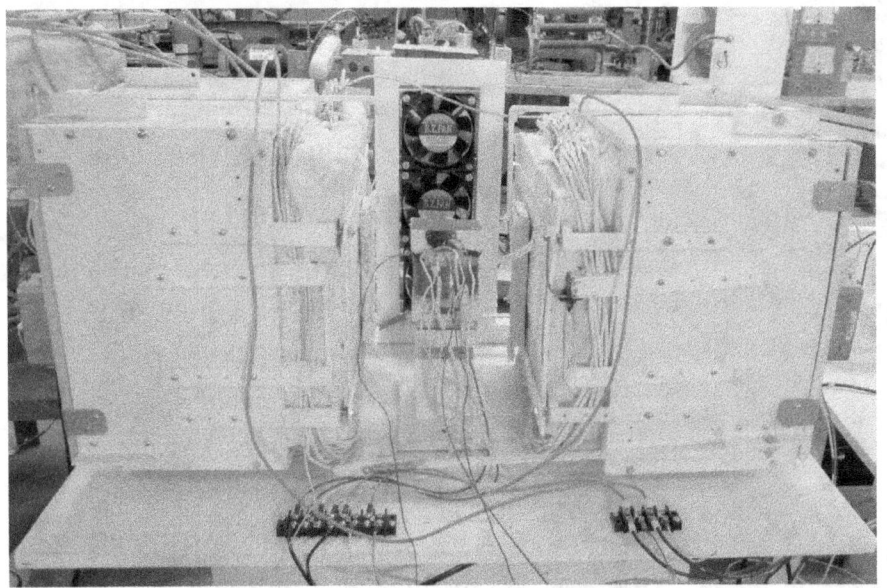

In this picture, the two heaters with the core laminations are in the 'pulled back' position. When pushed up tight against the transformer, four springs keep the pressure on. Even so, when a current goes through the coil it does buzz loudly as the lines of force created by the transformer's alternating current works against the steel's magnetic hysteresis.

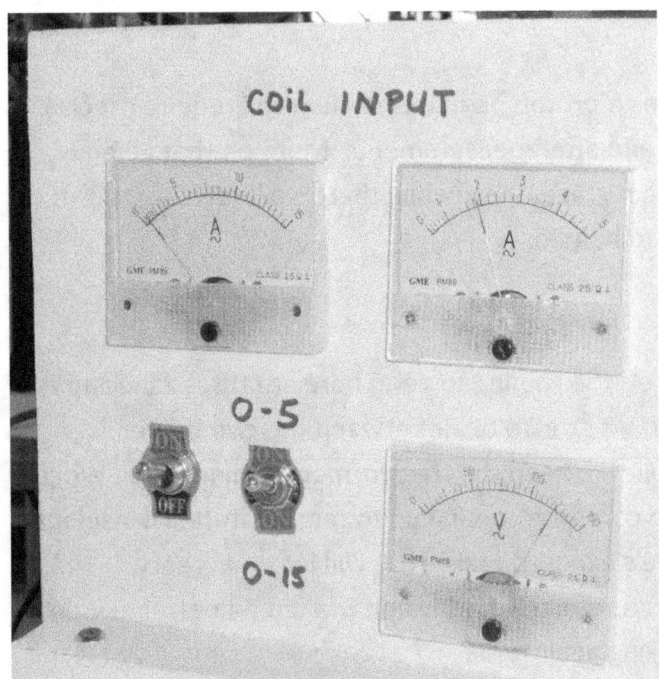

When the heaters are separated like that, the coil doesn't act much like a transformer. *(more like an electromagnet)* It draws 12.5 Amps at 120 Volts AC. That represents the current required for the coil to send all of its lines of force through the air.

Now, when the (cold) coil is pressed up against both (cold) core sides, that coil current drops to just 1.6 Amps. *(If it was still a perfect transformer and had not been cut apart and given mild steel sides that don't butt up perfectly, the idle current would be around 0.2 amps.)*

I'm showing it running with the core cold and the sides pressed against the coil. Note how I need two meters, 0 to 5 amps for when it's cold and a 0 to 15 for when the core gets hot, loses permittivity and the coil current rises above five amps.

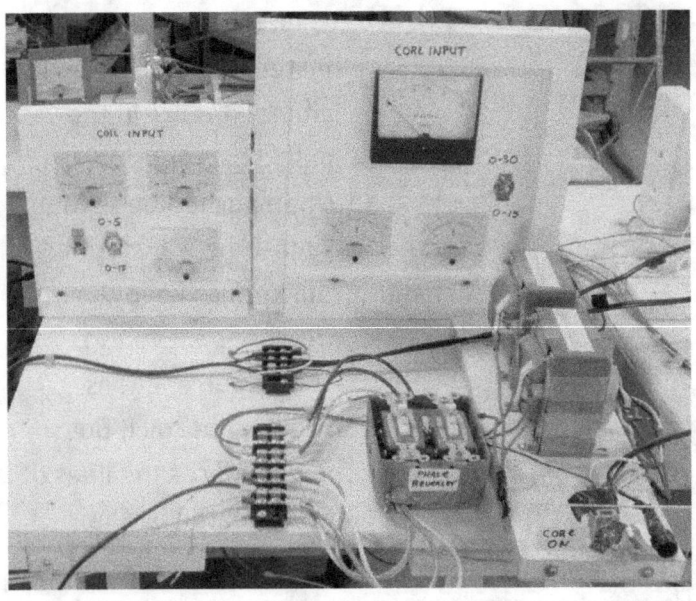

To measure the core input, I'm using this panel. It's the core power input table.

Those two transformers on the table change the 120 V line current to 13.5 V at a high amperage. Connected in parallel like that they can send 18 amps through both core sides. I assume the current will drop when the core is hot, so once again I needed two meters to get the full range with some accuracy. I used those ganged switches from the old prototype, so I can reverse the phase of the core current and try both ways.

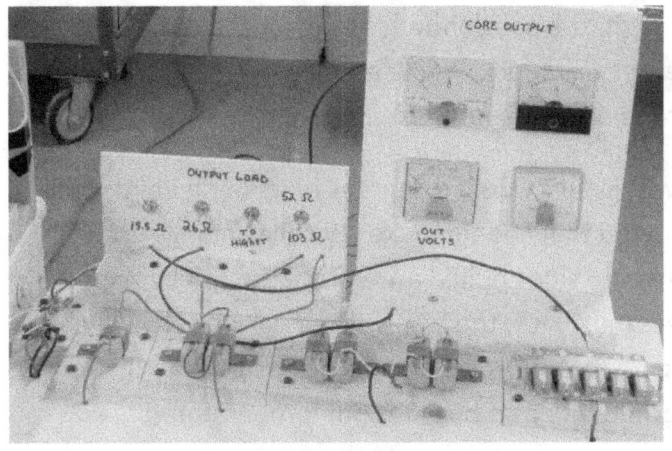

There is a secondary winding on my coil, so it acts like regular *(if not well-made)* transformer. When I put 120 V on the primary, the secondary winding shows a 40 V AC output. I send this to the output table where I can monitor the power coming out and watch for any changes between the core's cold state and after it's heated up.

With my simple equipment, I'm not sure if those changes will be most noticeable when looking at the voltage or the current.

To cover both conditions, I can choose between several loads. The lower ohms allow more current, and the higher resistive loads will make any voltage fluctuations more noticeable. I'll also connect my oscilloscope to this output.

After Trying

I felt sure adding extra insulation would allow the core to reach 1400°F. Seemingly I was wrong. Neither side can get over 1200°F. Those 300-Watt heaters simply don't create enough heat. I'm limited because the store-bought heaters are only rated for a top end of 1800°F. Even after reaching that temperature, they can't send enough heat down the copper rods to overcome my various heat-leaks. Those metal brackets suck so much heat away the temperature at the center of the core doesn't approach iron's Curie Point.

I've done some runs. At least that gave me a chance to check out the electric side. Yeah, once I push the core up against the coil, the time I get to work before the enamel winding grows too hot <u>is less</u> than a minute. Then I need to back the hot cores off and let it cool down. It's a strange way to experiment, but I didn't have to build any more crazy plumbing systems.

At 1200° the core's resistance rises, so the 17 amps that went through it when it was cold, drops to 10 at the higher temperature. More telling: My coil, as it acts like a transformer, draws 1.6 amps with both cold core sides pressed tight against it. *(When there is an insignificant draw on the secondary.)*

After the core has become as hot as I can get it, the coil continues to draw the <u>same</u> 1.6 amps. *(Remember the permittivity graph I drew at the beginning? It's only when the temperature gets close to the Curie Point the metal starts to reject those lines of force.)*

So, I conclude that when the core comes within a few dozen degrees of its Curie Point, only then will the metal's permittivity start to drop. And that should make my coil windings draw more current. In testing, I found that with both core sides pulled away from the coil by ¼ inch, the coil drew over six amps. When the core was fully back, the transformer, lacking any way but the surrounding air to complete its magnetic path, drew twelve amps. This assures me that even at 1200° the core retains close to the same permittivity as it has at room temperature.

Should I ever get the core close to its Curie point, it should show up because the coil will suddenly draw current as if it had an air gap. If the ferromagnetic effect works, when I send a current though the core, the amperage drawn by the coil should show a drop.

It's disappointing my tricks to get the temperature up to 1400°F didn't work. I felt sure by increasing the insulation I could get those last two hundred degrees. The only way I can work my way out of this dilemma is to buy more powerful heaters. Unfortunately, heaters don't come in every size. I'll not find any to directly replace the ones I'm using now. I'll have to build larger boxes to hold new ones.

Rebuilding the heaters.

I've found a source for more powerful heaters. They're larger. (6X6 inches rather than 4X6) That means both heater-core assemblies need to be disassembled and rebuilt. This is going to take a while. And even when I've done, I doubt it'll be the last word on whether manipulating the Curie Point can make an efficient generator. I've only tested one or two parameters, and those were the simple ones. Those I could do in my shop with not-very-advanced equipment and resources.

Disassembly

Ceramic wool insulation works well. Here's the inside of a heater box after several runs. While the core temperature never got much over 1200°F, I pushed those heaters close to 1900°F several times. Getting them that hot took about four to five hours, so the insulation got a good work-out.

What I did discover was those copper rods, which I'd used to channel the heat from inside the furnace chamber to the core laminations, worked against me. Getting copper red-hot in air forms various oxides which flake off and fall onto the bottom of the furnace. The layer of black flakes then insulates the heating elements. *(All the rods appeared much thinner, a few more runs and there wouldn't have been any left of them at all.)* I'm sure this coating reduced the heater's power and made them less efficient. Next is a picture of the furnace's inside after I got it open.

When I build my new furnace, I'll nickel-plate the copper rods so they don't disintegrate as quickly. I can't take the copper rods out of the core laminations without completely disassembling them. That means I'll need to sand-blast each lamination clean of old insulation and redo everything.

Here's a rundown of the next heater assemblies.

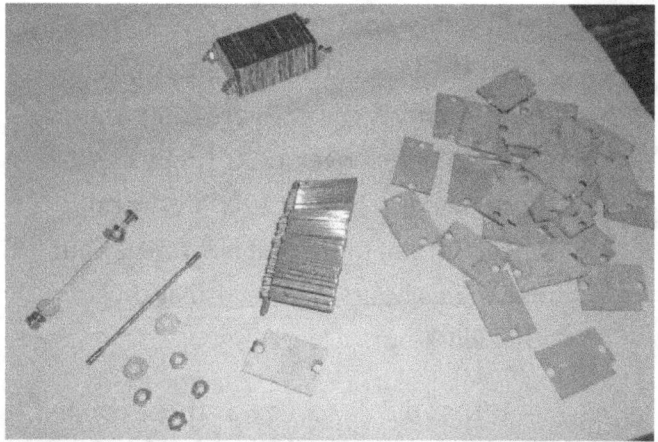

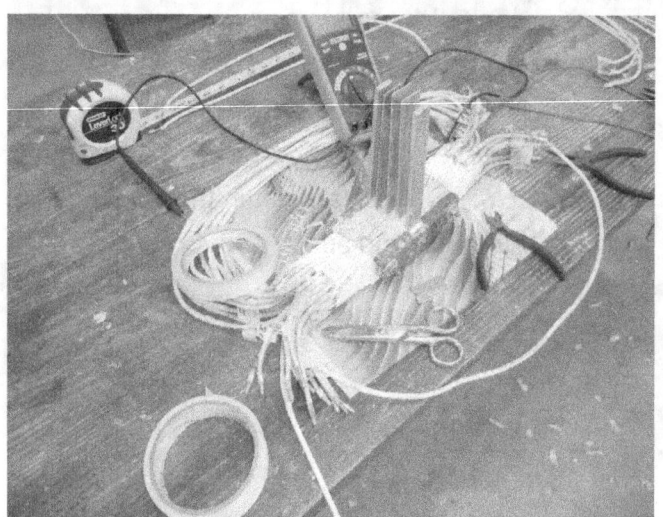

First, the bigger heaters needed bigger insulation boxes, as I want at least four inches of ceramic wool between the heater sides and the wood of the shroud box.

Taking the core assemblies apart meant sand-blasting each lamination and rebuilding them with new insulation. I also found those little end pieces I'd made to increase the insulation fell apart when I touched them. I decided to make larger and more sturdy ones. Instead of glue I used threaded rods *(wrapped with insulation)* to hold the laminations together.

A lot of machining gave me all those little bits. Assembled they make one-inch long core extenders.

I won't go into how much fixing and repairing I found necessary, but I was able to re-use most of the loop wiring from the first try. It was a slow process.

Here's putting it back together. All the laminations were cleaned, sandblasted, re-insulated and reassembled with nickel-plated copper rods. As I connect each lamination to the next, it's necessary to check the continuity and make sure no lamination shorts out to the one next to it.

The redone cores went into the new heater boxes, insulation put around them and the faceplate bolted into place:

The Rebuilt MMF Coil

Because on my first try, the magnetizing coil windings really did heat up too quickly, I've decided to do something about keeping them cooler and give myself more working time. I placed an aluminum base under the coil and water tubes in an aluminum heat sink on the top. In addition, I added two thermoelectric coolers on the coil.

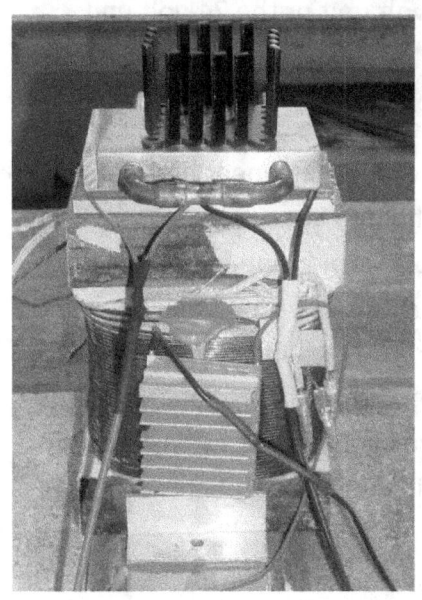

Now, thermoelectric coolers are not the best things. Besides pulling heat out of the core, they also generate extra heat. That means, after you turn off their power, considerable heat flows backward into the thing you're trying to keep cool. Once they're on, it's best to leave them on.

I'm thinking of using an "idle current" setting for the coolers. They cool best at 12 volts, but they'll work just enough at 6 volts to prevent the heat they've pushed up from flowing back into the transformer.

Here's another view with the better MMF coil placed between the heaters. Not all the wiring or cooling tubes are in place yet. It'll look a bit confusing once they're hooked up.

Meanwhile, my *(expensive)* new exciter transformer arrived. The ones I used last time, made from two 13 Volt transformers connected in parallel, didn't send as much current as I wanted through the core once they were hot. *(And I never did get the core all the way up to temperature.)*

I bought this larger unit. It puts out 24 volts and can go all the way to 40 amps, although I don't think I'll get that much current to flow when the cores are hot. I decided to use both current sources, *(since I have them made)* with a crossover switch to let me chose between the 13- volt unit or the 24-volt unit.

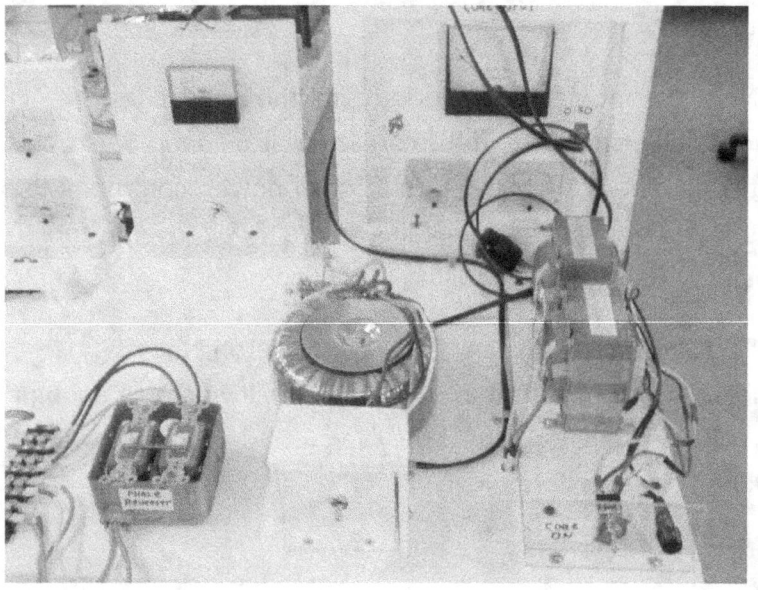

Note this new power transformer is a toroid. Now that some smart engineer has figured out how to get copper wires to wind onto a toroid shape, they are just about the only kind of high power units you can buy. Remember, the doughnut configuration is the most efficient shape for a transformer—but it's the hardest type to assemble.

Final Assembly

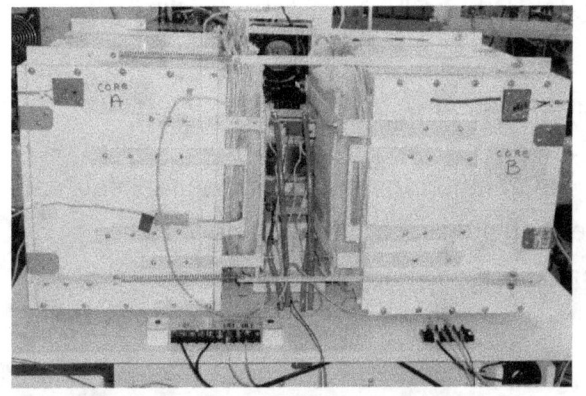

It took some time—and a lot happened to delay me—but here's an overview before I crank it up, beginning with the two core boxes pressed against the coil.

And here is the shroud that covers the exposed ends of the core while it is pulled back and coming up to temperature.

Here's my improved power input table. It now has two current sources for the core current, low and high. The MMF coil is still run from 120v house current. I'm using the DC power supply to run the thermoelectric cooler, which I'm hope will keep the coil cooler longer and let me make more measurements.

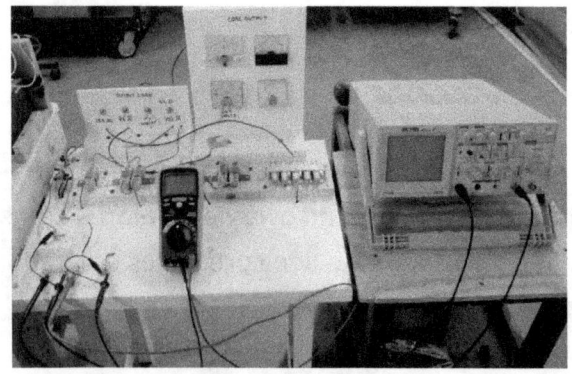

The secondary coil output table, with its load resistors. Alongside is my oscilloscope

The scope monitors the input on the MMF coil and the output of the core.

Remember, the changing magnetic flux and the changing electric current need to be in phase with each other. The simplest way is to keep the number of transformers in each circuit identical. When running, this is how they look.

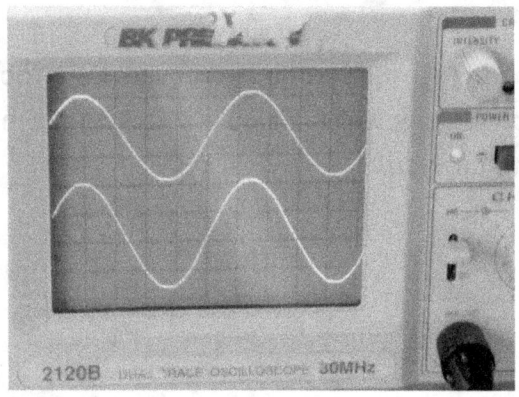
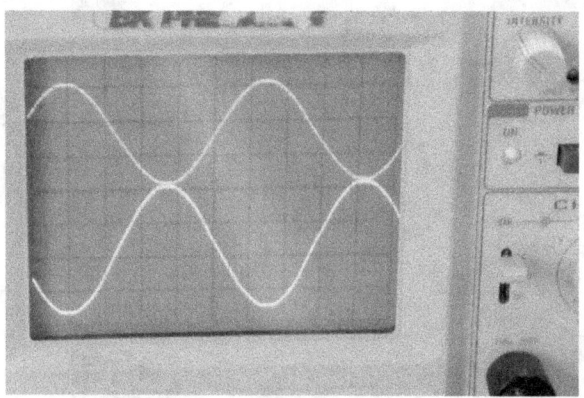

I have a phase reverser on the core input so there are two ways the phases can interact with each other. That's why I'm showing two views of the scope.

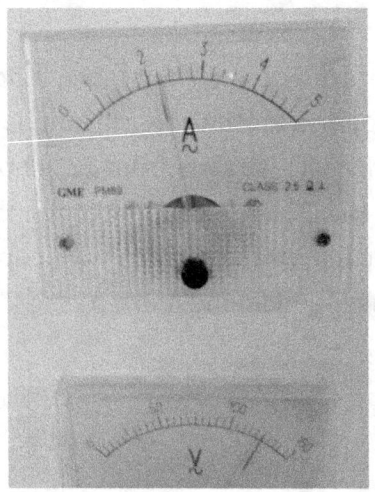

At room temperature, here's the current the MMF coil pulls at 120V, with the cores butted tight against it.

A little higher than before, but the new extenders are longer than the first ones I used. Did I mention that mild steel is not the best for making transformers?

As the core gets hotter, I'll watch this current rise. When the core gets to around 1400°F the current should be higher. Then, when I activate the core current, the MMF current should drop back down. *(Suggesting the core has regained some permittivity.)*

END

It blew up!

Then it caught fire.

I got less than 30 seconds to play around, but in that time, with the temperature fluctuating so wildly, I couldn't detect any change when the cores were close to 1400°F.

It took far longer, several hours, for the temperature to rise, but I finally cajoled both cores to 1400°F. Then I jumped up and ran around like crazy removing the end shields, shoving the two boxes against the coil and connecting the pressure springs. Getting back to the control table I found the MMF coil now drew 5.2 Amps instead of the 2.6 amps it drew when it was cold—indicating the center part of the core had lost its magnetic properties. But when I sent a current through the core, *(both ways)* it remained at 5.2 amps. Then I looked at the core temperature thermocouples and they'd jumped beyond 1500°F. I was reaching for the heater switches thinking I'd let the core drop down to something below the Curie Point when heater "B" produced a shower of sparks. I hit the heater's power switch off, but then the wood shroud on heater "A" used the opportunity to burst into flames.

I can't say for sure what the temperature of the core really was when I made that short try. With the temperature jumping around like that a significant portion in the center could have gotten above the Curie Point before I applied the current. I'm sure that once the iron goes beyond its Curie Point, it can't drop back down. The central portion of the C-laminations could have well exceeded the temperature at which the effect could take place. With so little to go on, I can't tell.

One heater is dead. The shroud on the other caught fire, but it might be salvageable, I'll take it apart and see if there's anything I can do. Maybe I'll get lucky. But it'll take quite some repairing to get to the point where I can give it another try. And, in all honestly, I'm not sure I should bother. It was pretty much around the temperature where something should have happened. I now wonder if my idea of placing the core laminations parallel to the flux was incorrect. I knew it wouldn't be as effective, but I thought I'd see some reaction.

Later, dismantling

I discovered my error after I took one of the core heater set-ups apart. Another DUH! moment. Even those more powerful heaters weren't getting the iron C-laminations *(which gets their heat from those copper rods)* to 1400°F, and I forgot the melting point of copper is 1981°F.

Yeah, in my desperation to finally get some results, I let the inside of the heater chambers go above that point. It's probable the core thermocouple wasn't able to get any kind of accurate reading with so much breaking down. Plus, with the copper strips that hot, it's probable the centers of the cores were already above their Curie Point even though part of the core remained below that temperature. Meaning I had a substantial air gap that wouldn't go away.

The overheated copper rods began melting and molten copper dropped down onto the lower heater coils. There it flowed across the heater strips, burned through the ceramic cement and crazy things began happening.

There are four heaters in all, and on both sides, I've ruined the two bottom ones. That gives me two heaters *(the top ones)* that still work. I could rebuild one heater-core assembly and try it with just one side. But the copper heat-transfer strips have melted and will need to be replaced and that involves taking the cores completely apart yet again. *(A very long process.)*

This picture shows the melted rods.

And the bottom of the heater: That's a puddle of solidified copper shorting out the heather coils.

Wrapping it up.

I never got "I'm sure it doesn't work" results, so I suppose I'll keep plodding. Crazy means doing the same thing over and over and expecting different results. I qualify on that count. In my defense, each prototype had recognizable flaws, which I tried to correct with the next go-around.

I remain convinced it's possible to manipulate the Curie Point and create a whole new kind of efficient electric generator. My simple ideas could use refining from those with real engineering degrees. I suppose this try-and-fail approach harkens back to a time when people played with pans of mercury and bits of metal stuck into lemons.

The world has moved on from that time. I was sorta working on the idea this little bit got skipped over and now, in an era of integrated circuits and cell phones, nobody gives much thought to eighteenth-century probing.

Certainly, the world could use a straightforward and cheap way of harnessing solar power, one the oil and corporate lobbies can't dismiss as easily as they have with other not-in-their-interest ideas. If this book has any message, it's that we should be willing to examine off-beat ways of exploiting solar energy. Burning carbon is just too costly and we can't continue to do so for much longer anyway.

So, I haven't given up. This started out to give me something to work on while taking care of my wife. It did keep me sane, *(mostly)* so I don't regret doing it. Everyone needs a hobby, and I'm comfortable tinkering with it. Right now, I need to come up with a way of controlling the temperature more precisely. It means starting over, so whether I'll get another prototype built and tested is up in the air.

Things happen when they're supposed to happen. There will be no easy way out of the current energy/economic crisis until this generation accepts its responsibility to the next.

Like it or not, we're all in this together.

PART TWO

CHAPTER 1

Another try

After the inglorious demise of my last prototype, I spent a long time going back to the books and reviewing the underlying physics and reconsidering every aspect of my idea. I suppose I hoped to find some flaw that would assure me this idea held no merit and I could abandon it and get on with my life. However, several months later all I can say for sure is that there is <u>nothing</u> yet understood that makes it impossible to electrically affect the Curie Point of a ferromagnetic material. My way of doing it might be wrong, or more likely, using my crude prototypes I've never gotten all the conditions exactly right, but everything in physics suggests Curie Point manipulation via electrical stress must be achievable.

Out of desperation, I've decided to build one of my more radical ideas. Admittedly, it's not a great idea. I'd previously rejected it as having too many different ways in which it could fail. But it is a way to overcome my endless problem of only being able to work with mild steel and its over-high Curie Point. So, instead of trying to find yet another ridiculous way to keep the field coils cool while heating the core, I'm planning to heat up the entire prototype—the ferromagnetic core and the field coil !

This will, by necessity, be a simple design—just an inductor with an iron core I can put into a high-temperature oven. Remember how the inductance of a coil changes when iron is inserted into its core? Same idea. The core, when heated to almost its Curie Point will possess little permittivity and the inductor windings will then act as if they have an air core. However, when a changing current runs through that core, it will become more magnetic and the inductor's overall reactance will increase causing the current to drop. My intent it to monitor the inductor's current, and see if during those moments when the core is active, the current is reduced.

Yes, this is a dumb design. It will have nothing to do with a working generator. Only by using a core with a Curie Point lower than the field magnets can a real generator be designed. But if this idea shows promise, then it'd be worth developing such metals.

Essentially, my plan is to compare the current drawn by two inductors. One with a ferromagnetic core and one without:

Higher inductance means lower current, and there will be a higher current with lower inductance.

If I were a rich research institution, I would make two inductors identical in every way except for the core material and heat them together in the same furnace. But I'm not, so I'll only make the iron one and heat it up to close to its Curie Point where it mostly resembles the air-core inductor and then watch for any change when I apply a core current. Hopefully, it'll show signs of reverting to acting like the iron-core inductor again.

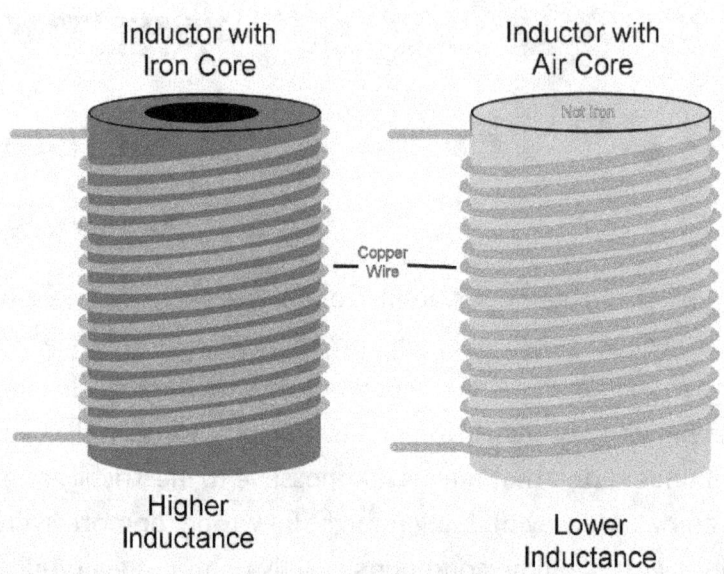

Like all my simple designs, this one will be especially difficult *(slow)* to make. Decent Inductors are made with magnet wire—copper coated with a thin insulating varnish that prevents the current from jumping sideways from one turn to the next. As previously noted, varnish can't get above 200°C or it starts to decompose. The residue of burned varnish is mostly carbon which is conductive. The insulation on the copper winding of this inductor will need to withstand 760°C. Making an inductor able to survive and not short out at such a high temperature will be the hardest part. In addition, with the materials I have available, the wire's insulation will be thick—meaning there won't be many turns per unit area. As an inductor, it won't have a high reactance causing it to draw a significant idle current.

Measuring, what will probably be only a small change in a hefty current, will be a challenge. Plus, I'll have to consider the natural increase in resistance of the copper winding as it gets hotter. My plan at this point is to hold the temperature constant just below the Curie Point and then turn the core current on and off to reveal the effect.

Beginning the work

First, I'll need a new core. As I still had my previously used core-winding set-up lying around, I got it going again. This time I'm using it to create <u>four</u> coiled pies. These will be smaller and made from mild steel rods 1 and ½ inches wide, 1/16th inch think and 6 feet long. Stacked together the pies make a core 6 inches high and 3 ½ inches in diameter with a 2-inch hollow center.

Here is one at the start of the wind. Note the ceramic cloth sandwiched between the layers as the winding progresses. That is used for electrical insulation so the current doesn't jump between spirals.

After one core section has been wound, removed from the winder and annealed in the oven so it takes a set, it looks like this:

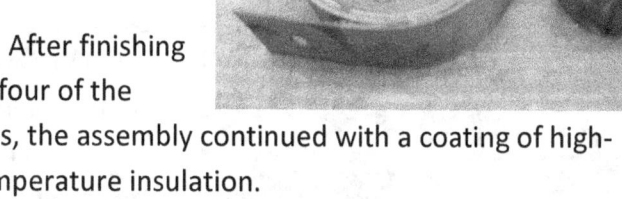

The smaller metal 'donut' is used for winding. After this step, I cut off the end tab of the spiral and welded on the contacts so one core pie can electrically connect to the next one in the stack.

After finishing all four of the pies, the assembly continued with a coating of high-temperature insulation.

This picture shows the pies stacked up and you can see the top electric tab. The other tab is on the bottom. As the cores are only held together by the welds that allow current to flow from one pie to the next, they need an outside brace. That's what those brackets and bolts in the next picture are for.

However, I can't leave external bolts in place while applying the copper wire. Plus, I'll need a way to turn the hollow-centered core. At first I thought I could use that wooden insert with the square bar behind the core during the coil winding process, but quickly realized wood would not have the strength needed. I discarded the wood and built a metal turning assembly which grips the unit more strongly when the outside bolts are removed. The square rod down the center will be the part that turns and makes the whole unit rotate.

Here you can see how the center insert will both hold it together and transfer the rotation from the square bar.

And now begins the long part. I'll use bare copper wire and squish high-temperature cement insulation between each strand to keep them from touching. By working in layers, a sheet of insulation can be placed between each layer. This cement, although good for high-temperatures, needs time and exposure so it dries and sets properly.

To make my plans more understandable, I've draw a picture looking at the inductor as if it were cut in half lengthwise, so the ends of the wires appear as round circles.

The metal core is heated from its interior cavity. Wrapping around the core is a thin layer of insulation. In this case, it's primarily electrical insulation to prevent shorts. Then I lay down my copper wire with a gap between each turn into which I force high-temperature cement. Once one layer is finished and dry, I cover it with ceramic cloth impregnated with insulating cerement before putting on the next layer of wire. Looking at an enlarged cross section of the top half:

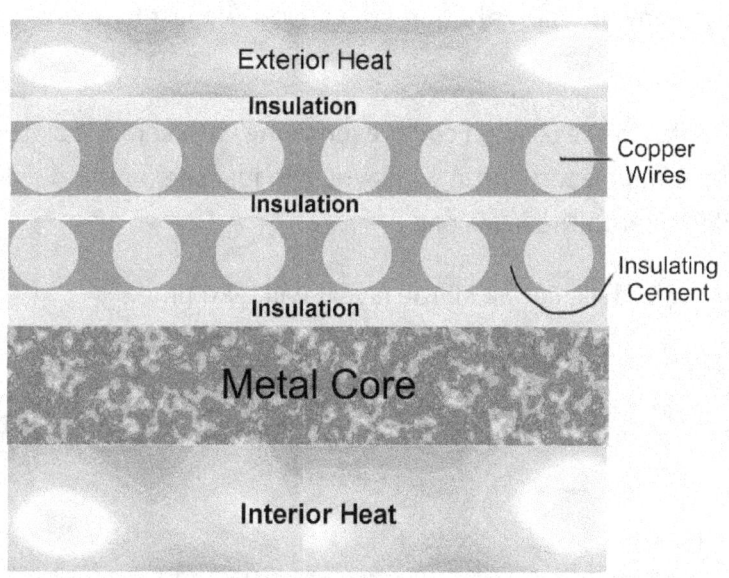

I only show two layers of wire, but there will be more layers before I'm finished. To make sure I maintained a gap between the copper strands, I wound the copper side by side with a string. Once a layer was complete, I removed the string and pressed the cement in the gap it left.

What could prove to be a fatal flaw in this prototype comes from a metal's tendency to expand when heated. The long coil of copper wire will increase in length as the temperature rises. *(It'll expand in all directions but lengthwise will be the problem.)*

Here is the coil in the process of being wound. You can see the string I'm using to maintain a gap between the strands of copper.

I'm trying to build in an expansion space and flexible joint at both ends of each layer in the hopes that doing so will prevent the whole unit from blowing up like a balloon when it's in the furnace. The expansion joints are a compromise, but I haven't been able to come up with a good work-around.

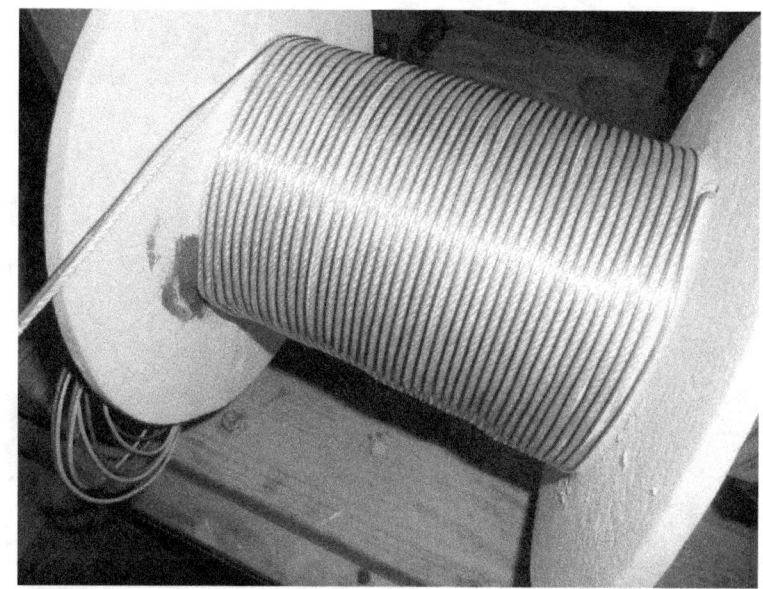

Interestingly, one way the ferromagnetic Curie Point has already been shifted is by clamping iron so tightly it can't expand when heated. While that wouldn't translate into a workable generator, it does show how stress on the linked dipoles has an effect. Now we just need to create that stress electrically.

After a layer of high-temp cement replaces the string, I cover everything with ceramic cloth hardened with cement before starting the next layer. Those wires you can just see on the outside come from the thermocouples imbedded next to the iron core.

This is a good place to stop, as winding and insulating all the layers will take time.

CHAPTER 2

Assembling the core.

Winding and insulating the core was a long process. Here it is after applying the last layer of copper wire. I geared the motor down so the coil turns around thirty RPMs, enabling me to guide the wire and string by hand.

After I finished winding the coils, I applied the final layer of ceramic cement to hold the wires in place. Then I replaced the bolts used to prevent the core from falling apart.

Although it's heavy, the wound core is not physically strong. Before removing it from the winder I needed to make a holder for it.

The iron holder is conductive. I took the precaution of gluing insulating ceramic extensions to the legs. That way, when everything is red hot, there will be less chance of the core current finding an alternative path through the metal. At high temperatures, electric current doesn't act the same way as it does at room temperature.

The holder does double duty in that it also supports my inner heater.

I'm using a store-bought stick heater, which can put out 2000 watts. When everything is assembled, it will pass though the hollow center of the core.

Here is the core sitting on its holder.

Note the ceramic washers on the long bolts. I used those washers to insulate the long bolts from the metal braces.

Next, the core needs a heating chamber built around it. In this picture, you see the core, in its holder, sitting on insulation and firebrick.

I'm reusing those two heaters which survived my last prototype. They will work in addition to the heater which runs down the center of the core. I've attached them to the sides at slight angles. If the ceramic cement holding them fails, they won't topple onto the core.

And the side heaters fit around the core like this.

The other two sides of the heater box I'll use to allow the various electrical wires connecting to the core and coil to extend beyond the hot zone.

There are heater wires, the core and coil wires, and the thermocouple wires, they all need to extend out of both sides of the hot box. Here is one side.

Both electric (wire exit) sides are installed. I'm using ceramic tubes to protect and insulate the wires and thermocouple probes inside the hot zone.

Here's a close up.

After this, the interesting bits start to disappear as I install the sides of heater box. The sides are made from a hard-ceramic board good for over 2000°F.

After its set, ceramic cement will hold everything together. A reinforced lid for the hot boxes will help keep it from falling apart.

With the inner box completed, I start making the outer box that will hold the ceramic wool against it. Ceramic wool provides the main insulation which allows the core to come up to temperature. This time, I'm using non-flammable metal to make the outer shroud.

I'm reusing the wool from the previous prototype. All the space around the hot box needs at least four inches. *(I'll remove the wood bracing before final assembly.)*

Another view after more wool is carefully layered into the gap between the hot box and the metal sides.

Once the final metal side is installed and insulated, more insulation goes over the hot box top.

I made a metal lid to reduce air-convection cooling.

With everything under the insulation and covered by the metal shroud, it just looks like a big box

Only the wires protrude from the box. The next stage involves connecting those wires, and thermocouples, to the electronic tables.

CHAPTER 3

The electronics.

Getting the support electronics correct has proved to be as hard as building the core.

I made several versions, and after the first heating session, needed to do quite a bit of revising.

Here's an overall picture of the latest version: The magnets are energized but the core heaters are not on.

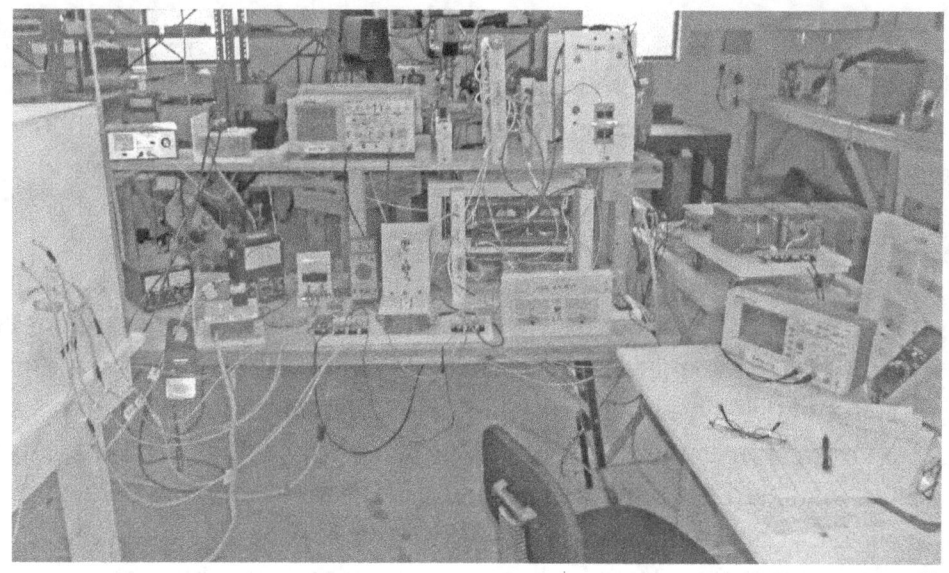

ASSEMBLING.

Providing AC power means carefully allocating the wall outlets in my shop. If I draw too much current from any single one, a breaker might trip at an inconvenient time.

First, there are the heaters. The 240V one goes on its own power line so isn't a problem. There are the two 120 volt heaters which I salvaged from my last prototype. Combined they draw around 13 amps, so they must have their own circuit breaker. Here's the simple schematics for the heaters:

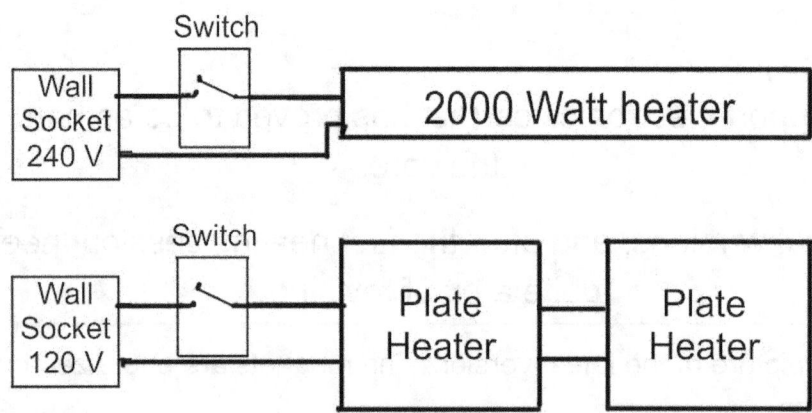

I'll manually control the temperature by turning the switches on and off while monitoring the readouts from several thermocouple thermometers. Remember, the 2000 watt heater is down the center of the core assembly and the two plate heaters are on the outside of the heat-resistant magnetizing coil. (MMF)

I intend for everything inside the insulating box to approach the Curie Point of iron. That's going to create a problem. The copper wire used to create the external magnetic field will expand with heat. I used expansion joints, but now suspect they won't be sufficient to prevent the coil from buckling and shorting out. No doubt, I'll not get as many attempts as I hoped to try different combinations of current and magnetic field. Now it's finished and sealed inside the insulated container, I'm wishing I'd made the coil with fewer turns of a heavier gage wire. Of course, that would have involved using a high magnetizing current, which had previously given me trouble. I've not even tried this prototype and I fear I know how it is going to fail. But I didn't go with high currents and fewer turns because it would make small changes in inductance hard to detect.

<u>Phasing is the critical</u>. I suspect one reason my previous prototypes didn't give any results was because I <u>never</u> got it correct.

The internal coil for the field magnetization has noticeable inductance. At 140 V AC, its reactance limits the current to just 4.5 amps. I made a step-up transformer which gives 240 V, so it sends almost twice that amount of current through the coil. The intent is to create a strong sinusoidal magnetic field inside the core metal.

Any inductance causes the current to lag the applied voltage. And the MMF inside the core is created by passing current through the coil windings. <u>Not by the voltage applied</u>. That means, to match the two phases, I needed to use an external phasing coil on the core current as the core's iron metal spirals have next to no inductance. Fortunately, I already had an external coil made and, with a bit of fiddling, adapted it to work here.

This external coil delays the phase of the <u>core current</u> so I can match its phase to the internal magnetizing coil. (<u>Which is fixed</u>.) A core's total inductance comes from both the number of turns (reactance) and the <u>overall resistance</u> of its circuit. To make small adjustments and get the internal coil and the external coil in phase, I use external resistors. These have the added use of giving me a way to observe the current's phase on an oscilloscope and compare them both.

After some work getting the phasing of the two coils exactly right, I realized that, as the internal coil heated up, its copper wire's resistance would change. Meaning, an arrangement which would be in phase when cold would not remain so perfect once hot. To address this problem, I built a resistor bank. This allows me to make small changes to keep the two currents as close to being in phase as possible. At first, I thought four resistors would be enough. After the first heated run I found it wasn't the case and needed to add a second and then a third stack, squeezing the big power resistors in where I could. I also added more fans to keep them cool. With these high currents even these one and two ohm power resistors get hot. A switch panel lets me engage different resistors as needed. I ended up being able to change between three, four or seven ohms to restrict the core current.

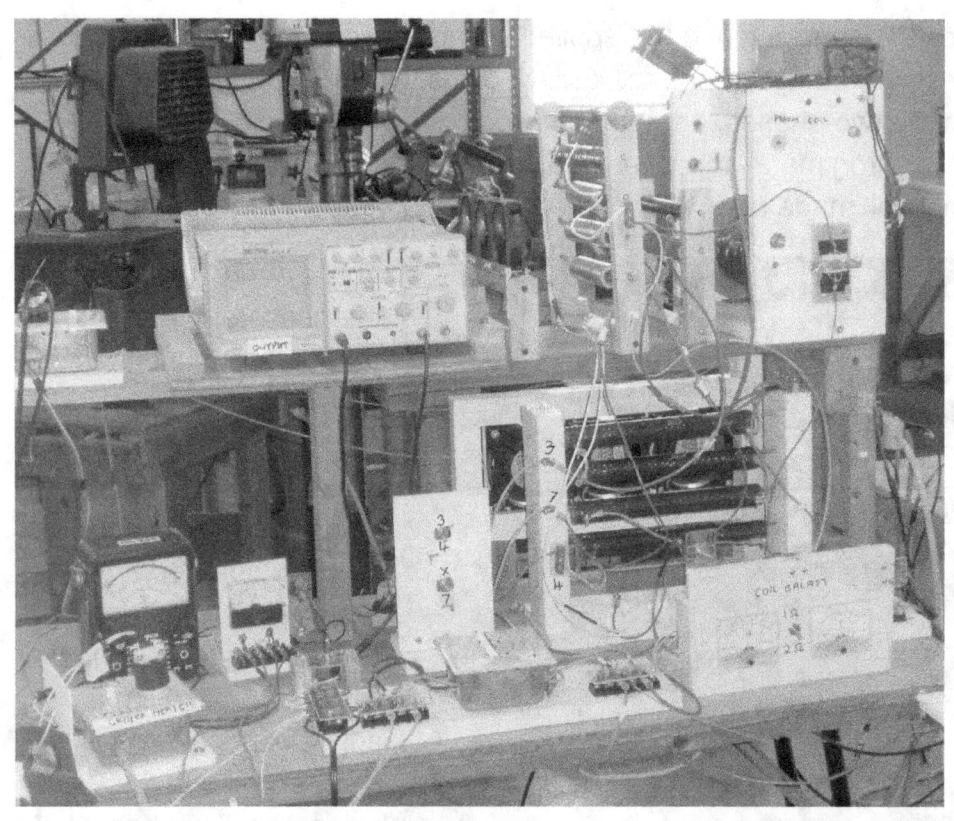

The **core** control panel has two sources of power. A normal 120 Volt souce coming from a wall socket and a 240 V source. To obtain 240 volts, *(and not as two 120 volt lines in opposite phase,)* I'm using two 1:2 transformers wired in parallel to get a high enough current. Note: to start these two big transformers together I need a starter resistor so a circuit breaker doesn't trip on initialization. Also, a grounding wire keeps the output <u>in phase</u> with power coming from the wall socket. *(That grounding wire carries a minuscule current of one or two µamps, yet it causes the phasing to stay the same as the input to the transformers instead of being 90° out of phase. Electronics sure are fascinating.)*

Here is the schematic of the core current:

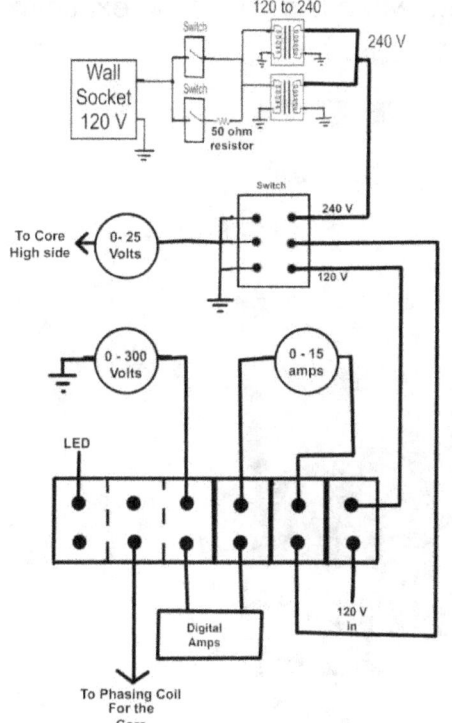

The core panel just shows the volts applied and the amps flowing through the metal spirals. A LED lets me know when current is flowing.

From the panel the current flows to the ballast inductor, through the balancing resistors, the phasing switch and finally through the core. Note how my resistor panel suffered a few modifications along the way.

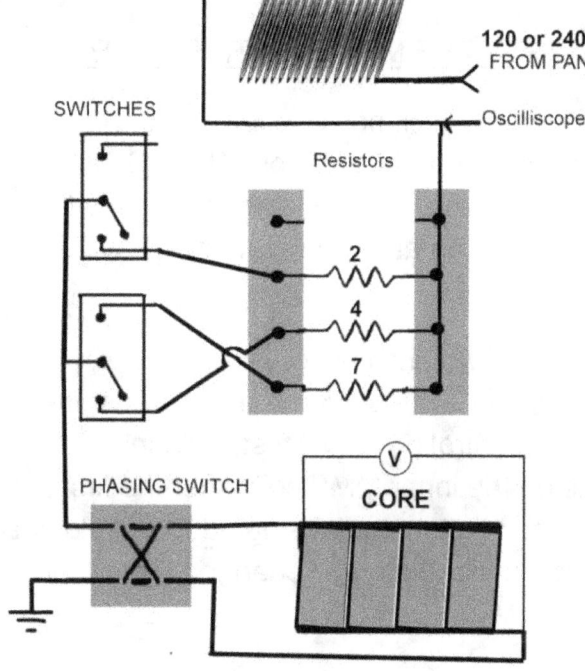

The picture is self-explanatory. But note, I placed the oscilloscope after the ballast coil but before the resistors. That's so I can monitor the phase of the current flowing through the core. The schematic only shows the path of the hot wire, but there is a neutral wire which goes to the phasing switch so I can try both phase relationships with the magnetizing current from the internal coil.

The phasing switch is the same as the one I previously used.

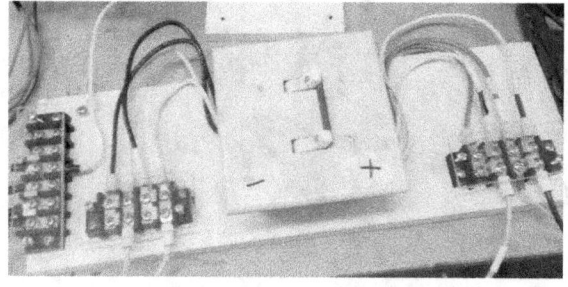

Here is the core current panel. Note the additional low voltage meter in the upper left corner. This meter gives me an indication of the AC voltage across the core. As the core is just a short iron path, there's not much voltage generated when it is cold, but I expect it to increase once up to temperature.

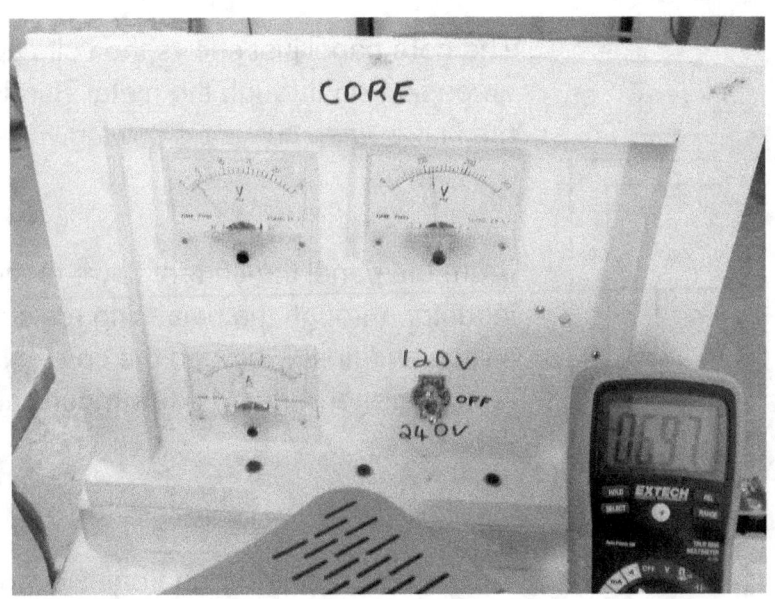

* * *

NEXT, THE INTERNAL COIL THAT PROVIEDS THE MMF FOR THE CORE.

The current that energizes the internal coil has its own panel, power supply and circuit. To give myself some flexibility, I can (now) switch between 140V AC or 110V AC inputs. These voltages come from an isolation transformer which slightly boosts or reduces the line voltage. (Another side grounding wire lets it act like an autotransformer and avoid phasing problems with the core current.)

Those are my two choices for powering the magnetizing coil. The power from the transformer goes to the control panel. At first, I planned to use 240 volts on the inner MMF coil. After the first hot run, I realized the cement-insulated coil could not withstand the vibration, so I changed the 240 V input to 110 V.

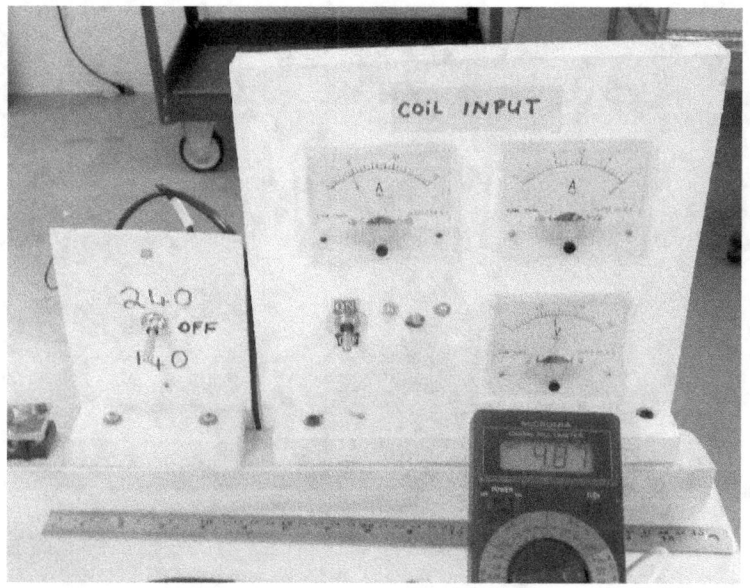

The inductance of the internal magnetizing coil remains <u>fixed</u>. But its copper wires will show an increased resistance at high temperatures, so, when the core is hot, it's necessary to have some flexibility for phasing adjustments. Besides, this circuit also needs an external ballast resistor so I can monitor the current on an oscilloscope and match its phase to the current in the core circuit.

The schematic looks like this:

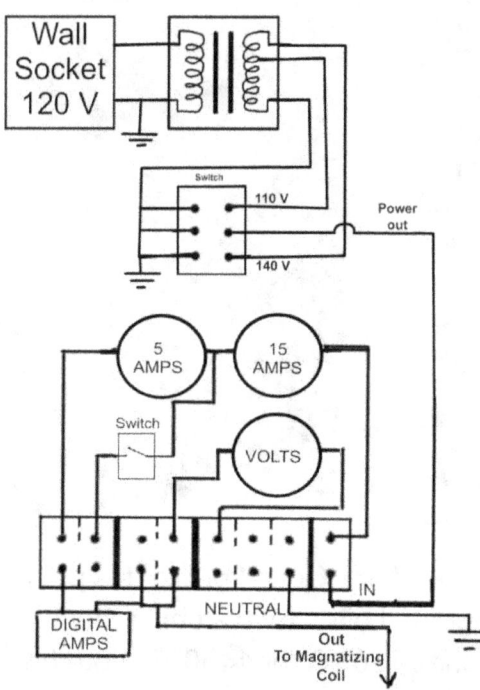

There are also three external resistors used in the magnetizing coil circuit: 0.03 ohm, one ohm and two ohms. These choices give me added phasing options. The bigger resistors are in the racks next to the core phasing coil. The small one, which I don't expect to use, is on the front side of the coil holder.

I realized watching oscilloscope or monitoring a digital meter with its last digits constantly flickering, would make it hard to spot small changes. I upgraded the coil ballast resistor panel and added analog meters to echo the oscilloscope. I'm of the opinion analog makes it easier to watch overall changes. Also, I used diodes to create a DC current. A DC meter uses a different mechanism than an AC meter, which I believe makes it more sensitive. Note how the absolute value of the DC milliamps is meaningless, I'm just watching the way the current changes.

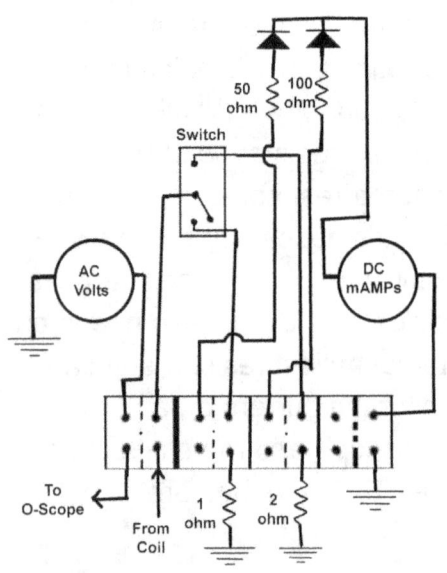
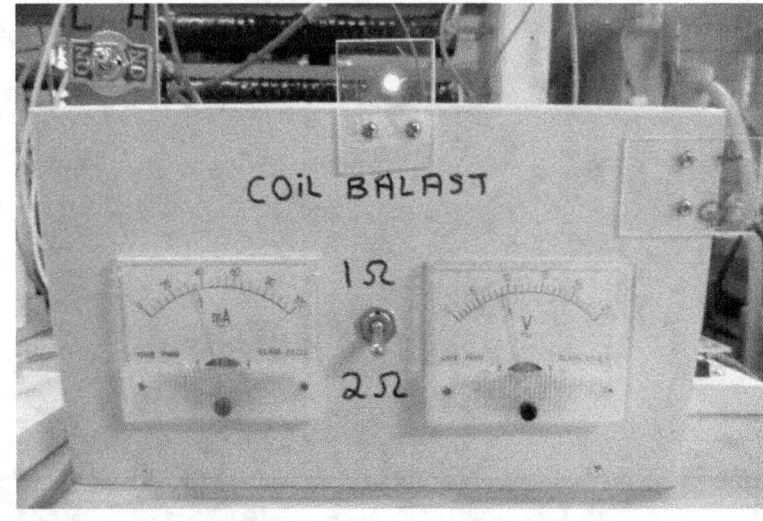

The wires for both the magnet and the core circuits enter the heated chamber. There's not much to see as it's all under the shroud and insulation.

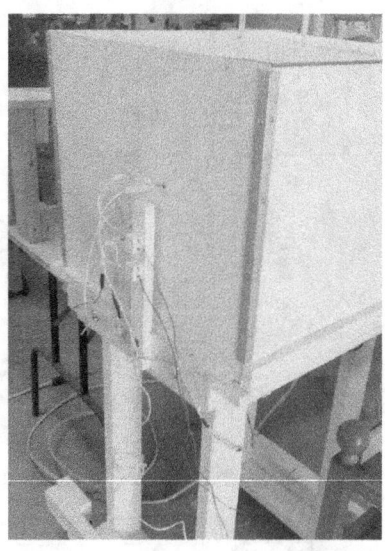

I have several thermocouples to keep track of the temperature. Once I get to this point, I wish I'd installed more back when I was building the internal bits. By watching these meters, I control the heating coils as the core approaches 760° C (or 1450°F) Note how in the photograph, these thermocouples, made for high temperatures, are not so accurate at room temperature. There are a lot of approximations in this design. As all of these thermocouples (probe ends) have been at room temperature for a long time, it would be encouraging if they all gave the same reading!

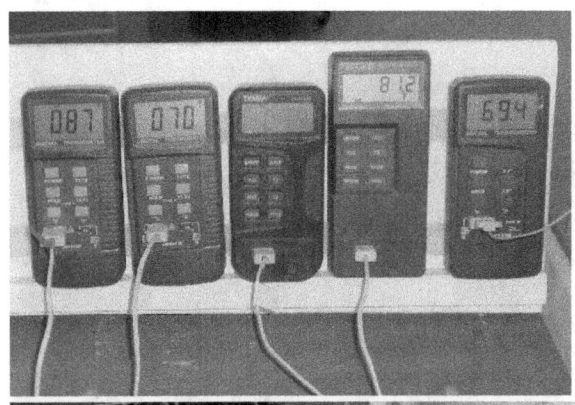

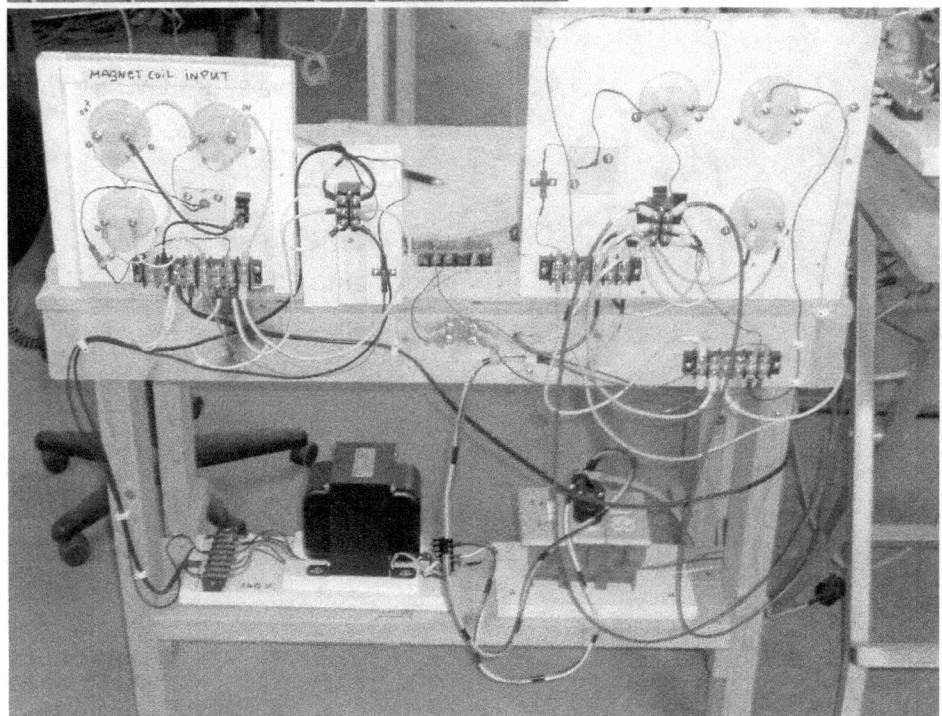

Here are the backs of the monitoring panels. It looks more complicated than it is. Note how I'm using lots of LEDs so I can tell if a circuit is hot or not.

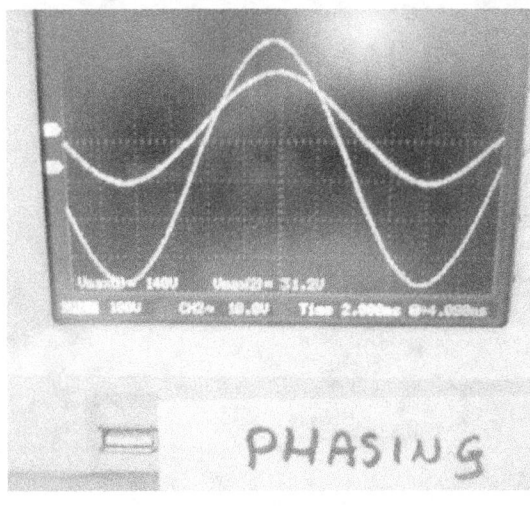

I use one duel trace oscilloscope to monitor the phasing. When displaying 60 cycles, one of the scope's screen divisions covers 21 degrees of the 360 degree cycle.

In this picture (cold) the two circuits are almost exactly in phase. I feel that should give plenty of the cycle where both the core current and the MMF are changing at the same time. Of course, it might not be that close when it's up to running temperature and the copper wires increase their resistance, but hopefully, by adjusting the external resistors I can keep them mostly in phase.

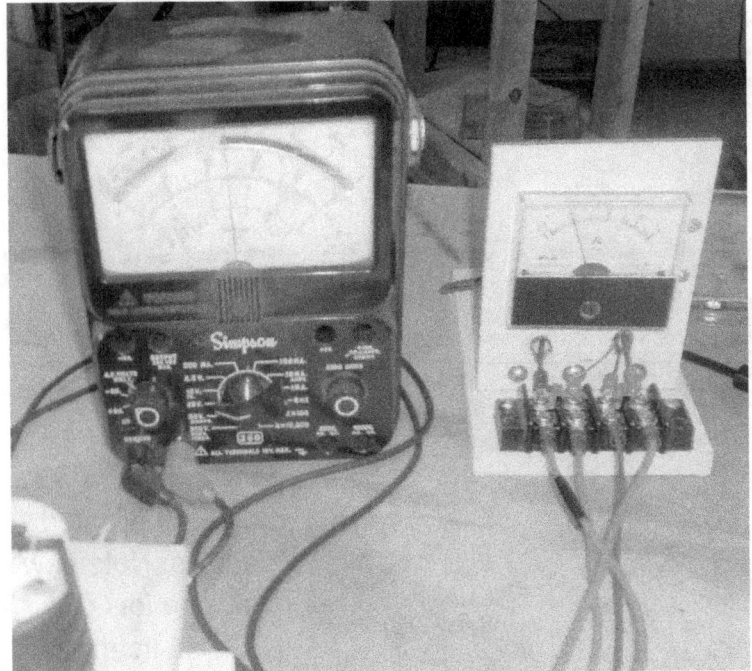

I do have a secondary winding on the internal coil to check the transformer-linked power produced by using a simple volt and amp rig. When the core is hot and non-magnetic, this secondary coil will have less transformer linkage, but its output should increase slightly when the core regains its magnetic properties.

However, being so close to the magnetizing coil, any small changes in output from this coil might not be noticeable.

I also monitor the secondary coil on an oscilloscope. It's the lower signal in this picture. It's referenced against the phase of the wall-socket sign wave which is my base standard.

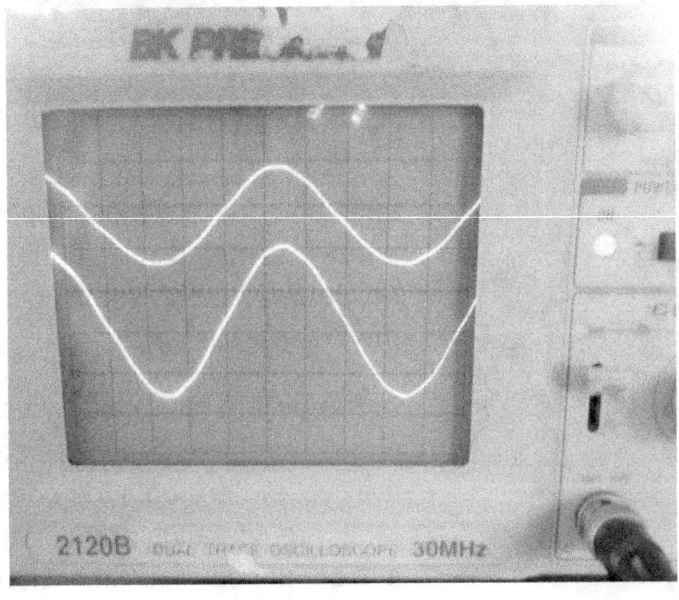

Chapter 4

Testing.

First try disappointments and then regrouping.

The first run didn't give me strong indications that it is possible to depress the Curie Point of a ferromagnetic material using electrical stress instead of mechanical stress or by reducing the temperature, but it didn't rule it out. Plus, I feel it gave me hints I'm on the right track. Unfortunately, several new problems prevented positive proof.

First, it took four hours to bring the core up to temperature. I wished I'd installed more thermocouples during the construction phase, as when the chamber heat approached 1400°F it became difficult to get a reliable reading of the metal core temperature. The outer and inner parts gave different readings and the insulated coil windings lagged behind making it even more confusing. I'm using grounded-style thermocouples, and even though I'd insulated them with ceramic, at high temperature the insulation leaked enough to cause problems.

As these thermocouples show, the inner hot zone and the air inside of the core is up to temperature. Yet the insulated magnetizing coil remains at 1300°F (the two readouts on the picture's left) making me unsure of what the temperature of the core iron, inside that coil copper was.

As I mentioned before, after initially applying current at this temperature, I realized the 240 Volt input to the coil surrounding the core wasn't useable. The resulting high current, and the weakening effect on the ceramic at such intense heat, created internal stress on the windings causing them to vibrate excessively. Too much vibration would break through the cement separating the windings and cause a short. I used only the lower amperage of the 140-volt input.

As I feared, the phasing did change as the copper of the inner MMF coil became more resistive. While I'd tried to anticipate this, my range of options proved insufficient. The phasing didn't remain as close as I'd hoped, although there was some part of the cycle where both the core and the MMF were changing at the same time. That's when I saw a fluctuating change on the output meters, plus it distorted the input coil current's sine wave. This was a distortion that wasn't present at just a slightly lower temperature.

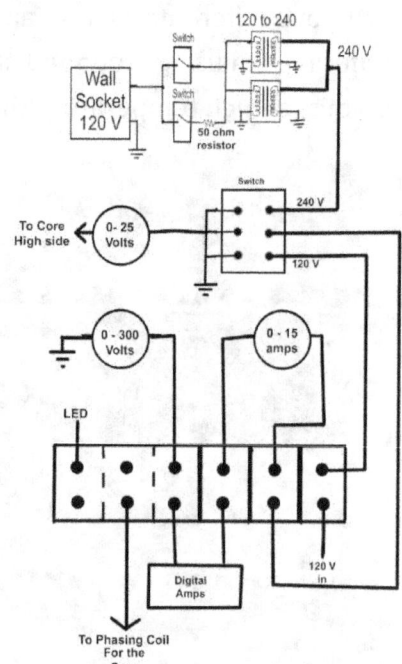

This insulation outgassed more than I'd experienced before. By the time the prototype reached 1400°F, I felt seriously dizzy. With that dangerous heat and the high currents I was using, I feared making a mistake and shut everything down before examining every possibility. Also, those ballast resistors began overheating and needed to cool down. The next day, after the core had cooled, my first task was to check the inner MMF coil. Fortunately, it didn't appear to have shorted while it was hot. It does hum more, probably to be expected from the gaps the AC vibration created while the insulation was hot. Like I said, I'll not use 240 volts on that MMF coil again, but the 140 V input continues to work.

When I try another run, I'll just use the 240 V input on the metal core pies. Also, I needed to rearrange and add to the core resistor banks as well as providing better air cooling for them. This schematic shows my latest configuration for sending current through the metal core.

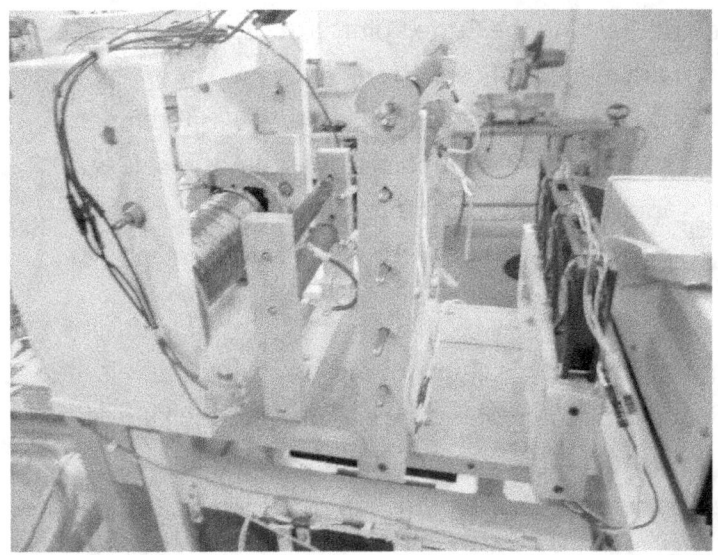

I needed to squeeze the rearranged ballast resistors into any available place. Note, I now have fans blowing cool air over them.

When the ballast is four ohms, and the voltage is 240, each resistor needs to dissipate 147 watts, and they are only rated for 100. However, I only apply current for short periods so they have time to cool.

If I'm careful.

The subsequent runs.

I had better luck after that first fiasco. I'd bought fans and ventilated the shop much better, plus it seemed as if most of the volatiles had burned off, so it wasn't as oppressive. Using only 140 V on the inner MMF coil, rather than 240, kept the vibration down. The core pies handled the higher current from the 240 supply well enough, although it made the ballast coil hum loudly and those resistors heated up amazingly quickly.

I still had trouble getting reliable temperature readings. If anything, measurements became worse than the first run, possibly because those grounded thermocouples now received even more stray voltage interference. I compensated by turning off the heaters during the heat-up phase and waited until the temperature gradients had mostly evened out. It took much longer to reach 1400° F this way, but I think I gained a better handle on the core temperature.

For the first time, I got a good indication of reaching the Curie Point. Just as core approached the critical temperature, the MMF coil began drawing significantly more current. Unfortunately, once it began to change, it proved impossible to slow it down. I suppose there was too much heat inertia in all that metal inside the hot zone. The thermocouples proved almost useless at this temperature, as they all gave different, rapidly-changing readings; some higher, some lower than the Curie Point. All I could do was kill the heaters, wait until the MMF coil's current dropped and try again.

I'm aware that after spending so much time and a large chunk of my life savings on this idea, I was primed to see what I wanted to see. So, I did try to be objective. I didn't get the kind of 'Wow' results a movie would portray, but it wasn't a complete 'no.' Mostly it was, 'there's something happening, but you need better instruments and a more sophisticated prototype to understand it.'

Just at that rather nebulous point where the iron began to show signs it was reaching its Curie Point by increasing the current through the MMF coil, the oscilloscope showed a small flattening on the core-current sine curve. That would suggest a second power source being added to it. At the same time, the core ballast meters flickered before returning to their original readings. But almost as soon as it began, it ended and the MMF coil current continued to increase as the full Curie effect took hold.

On my second or third try, I remembered to watch only the output monitor and saw the same small transient disturbance, which showed up more on the oscilloscope than the meters. Over several runs, *(Going back and forth over the point where the MMF current began to increase.)* I managed to observe similar events; sometimes more pronounced than others. Anytime I allowed the heat enter the core too quickly, if the event happened, I didn't catch it. Soon, of course, my poor, much-abused MMF coil showed signs it was finally getting internal shorts and no results after that could be trusted.

Two conclusions for future prototypes. First, my output monitor circuit drew only around 1 watt and the high currents and voltages inside the core and the MMF coil involved several hundred watts. Which meant the output had no effect on the system. That is, it didn't draw energy so there was no 'brake' to stop the heat inertia inside the core and coil from driving the iron toward its Curie Point. So, all I could expect is a quick flicker as the iron passed through the temperature where any real ferromagnetic effect would take place. With better thermocouples and a more sophisticated design, I might have been able to hold the heat in that critical zone and produce verifiable results. But I fear this prototype, with manually controlled heaters and way too much iron and copper, just wasn't up to the task.

Second, and more disturbing, is that whatever was going on, it only happened as the iron was <u>just starting</u> to approach its Curie Point. Once the iron's heat entered the downward slope of the BH curve, the stabilizing force of the electrical stress didn't have any effect on the dipoles. Obviously, this is a concern. A smaller shift in the Curie Point would mean a lower output even if it was used as a generator.

I'm hoping this problem was more a result of my system than a condition of the effect. Perhaps I should be happy I could see anything at all. As I noted at the start, this would be dependent on the electrical stress created inside the iron. The MMF coil was physically large, but only a little over 100 turns. No motor with such a poor power coil would be able to do real work. It makes sense that various combinations of magnetic flux and current would act in different ways. Like resonance, some combinations would be more effective than others, and my crude set-up was only able to show a vague hint of results at the weak beginning part of the slope.

Other considerations would be my using mild steel with its not-so-great permittivity. Low permittivity means the flux forced through the inside of the iron would not be so strong it could force all those lines into interacting with the dipoles. A core with a greater permittivity should operate lower on the BH curve and produce more power per cycle. The size and shape of the core will obviously play a part. This core only held <u>four pies</u> and they were 1 ½ inches wide. Plenty of room for the electric current to travel in a less than perpendicular path to the magnetizing field and such large conductors keep the overall current density low. This is definitely an area for more research.

Would 60 Hz be the best frequency? There are so many variables I haven't been able to address. I suppose I should be happy the prototype gave me a hint it might be possible to electrically change the Curie Point with electrical stress, just as already been shown to take place with mechanical stress.

Conclusion.

With my main MMF coil compromised, and no way to repair it without redoing everything, I must accept this prototype has run its course. It gave me a little hope, and once again showed me where I've made errors. I still have hope this could someday lead to an efficient way to harness the free and abundant power of the sun. But whether my efforts, or this book, will inspire anyone to examine the effect further, remains to be seen. Considering my age and health, I'm not sure I'll be able to continue this research much longer.

There are many ideas like this. It'd be great if one was to pay off. There are others, who like me, enjoy tinkering and trying out new things. If this book can inspire them and perhaps allow them to skip past some of the mistakes I've made, then I've not waisted my time. I do want to reiterate that nothing in any of the many physics books I've studied suggests this idea is impossible. In fact, it's just the opposite. Coupled with the hints this last prototype gave me, I feel I've established something worth pursuing. After all, the cost of finding out one way or the other would be far, far less than the money we spend on securing our carbon-based energy sources. Or dealing with that Elephant in the room: CO_2 induced Climate Change.

I'm hoping the idea we could tap into a fuel which doesn't threaten the climate and our way of life will have some small appeal.

CHAPTER 5

Post heating ideas

As my last prototype didn't provide undeniable proof sufficient to overcome the resistance of the energy barons, nor could I produce a neat media-friendly web video, I will continue my work for as long as I'm able. Pulling on all the mistakes I've made, and relying ever more on commercially available materials and parts, I've made tentative plans for my next venture. I'm putting them out hoping that others, with an interest in developing a way to free our world from its over-dependence on carbon-based energy, might want to take these ideas and work with them.

Once again, I don't plan to build a real generator. That would require materials and machining beyond what can be accomplished in a home shop. But this design should offer realistic proof of concept. It might even provide a nice video of the Curie Point changing. Its output should be an AC voltage at twice the line frequency.

This idea harkens back to the early work of Edison and others who used permanent magnets. It allows me to eliminate the flaw of heating up a long coil of wire to the point where its expansion causes problems. A small magnetizing coil of few turns would have little reactance and simplify the phasing problem, even though it will require a significant current. But, with the heavier wire's low resistance, just a small voltage will give a high current, so the input power would not be overly significant. I'm considering using just one copper-wire layer around a small (maybe just an inch in diameter) core.

To illustrate what I'm thinking, here is a top-down view so I can describe the parts necessary.

Note: I'm planning to build a long, curved DC electromagnet. I lack the tools necessary to bend steel, but it's possible to buy one-inch wide by one-eighth thick mild steel strips in 72 inch lengths. If I make a wooden jig, these strips are flexible enough for me to shape. I'll have to use eight of them to make a one-by-one inch curved bar.

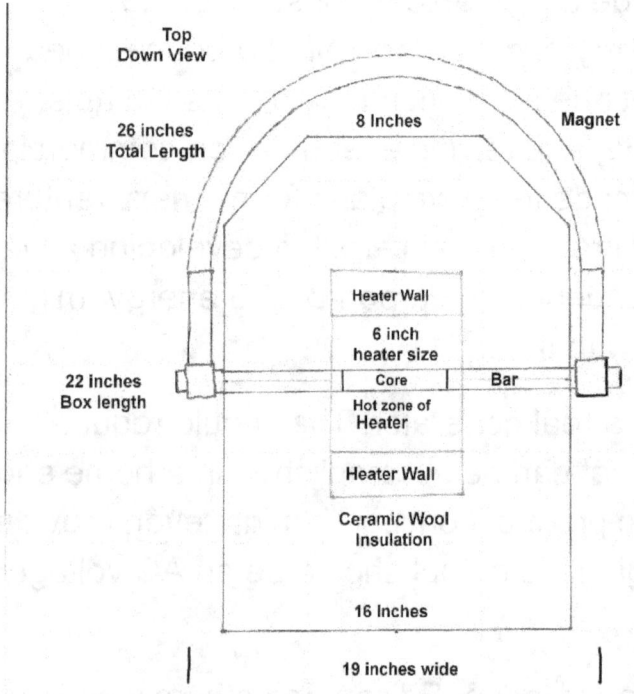

I'll use sheet-metal sides again, with lots of ceramic wool insulation to get the inner working parts up to temperature. I can buy 6 by 6 inch heaters. They're not especially powerful, only 300 watts each, as they're designed for long term use. But I'll have four of them working together. Two sides and a top and a bottom. The remaining two sides of the box will be open and available for the core and all the electronics to enter the hot zone.

The long thin horseshoe at the top of the picture is the DC powered electromagnet. It should create sufficient flux so a considerable amount will flow through the iron bar which runs through the hot center zone.

At the center of the bar, inside the heated area, I'll place the core pies. I do expect putting a mild steel bar near the heaters would allow too much heat to escape, so I'll resort to my trick of leaving them out until the core pies are almost up to temperature. That means, they'll need to be removable and replaced with ceramic inserts to keep most of the heat trapped in the hot zone during the long warm-up phase.

Next, I'll show a diagram of the inside of the hot zone made by placing four heaters in a rectangular tube.

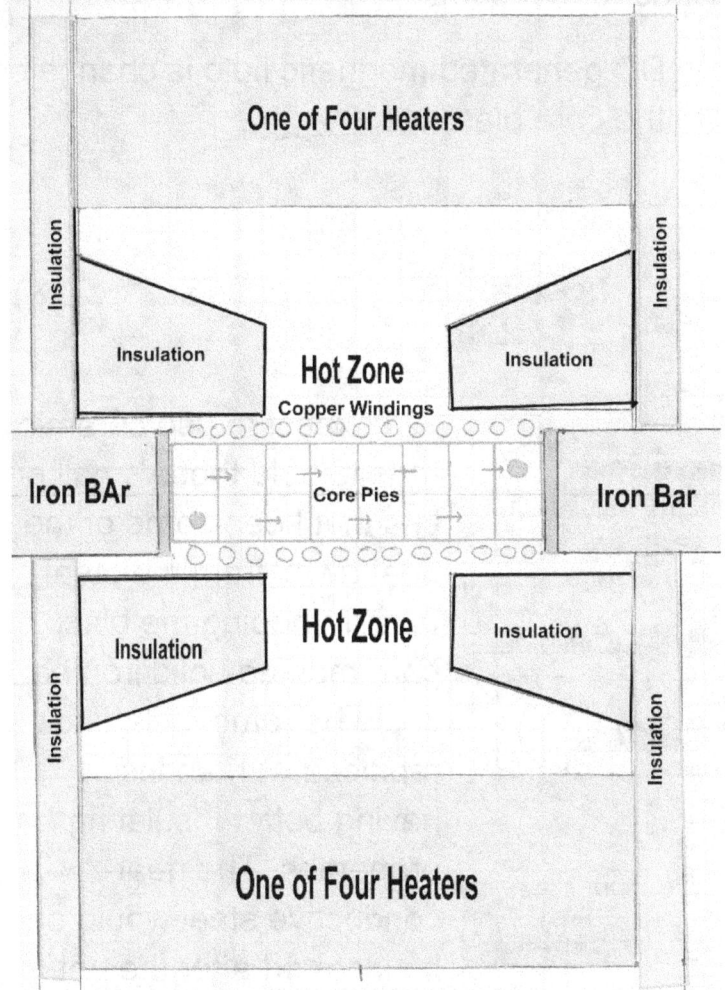

The heaters will be held in place by hard insulation. I'm considering using small insulation tabs inside the hot zone to concentrate the heat on the very center of the pies and hopefully not have too much leak into the mild steel bars.

This picture shows a core made from a one inch diameter steel tube with a ½ inch hole in its center. It's something I can buy, but I'll need to cut it up into ½ inch wide (or smaller) pies. Using a one inch bar, although rather small, is my choice because the maximum size a 5C collect can hold is one inch. A larger diameter tube would require better machine tools than I own. Also, drilling a hole larger than one inch *(for the joint where the DC magnet intersects the steel bar)* isn't going to be easy. There are a few solutions, but they'll cost money.

Around the pies, I plan to wrap a heavy copper winding to provide the changing magnetic flux. Current will come in on one side and exit on the other.

The pies will need their own electrical connections. All those wires, and the ends of the mild steel bars represent a heat leak the four heaters and their combined 1200 watts will need to overcome.

Here's a detail of the way the DC generated magnetic field is channeled into the core pies.

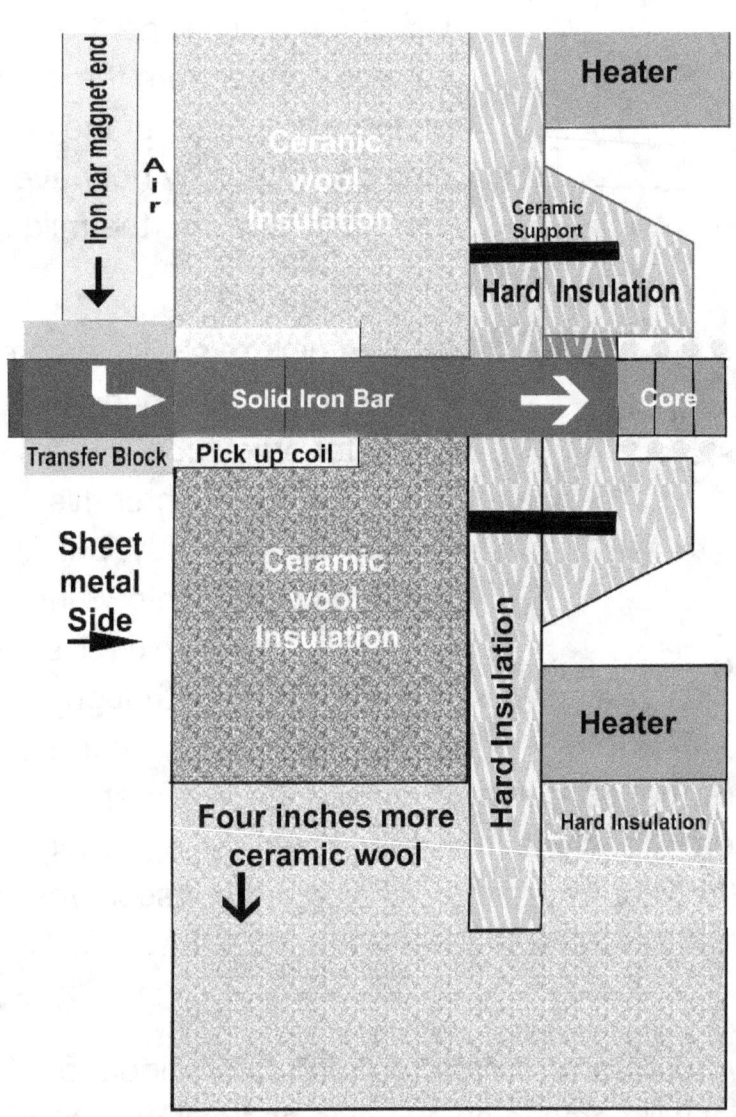

I expect to use ceramic rods to hold those small end bits that keep some of the heat away from the steel bar. I'm hoping the black part, marked solid iron bar, could be removable and replaced with an insert having better insulating properties. The heat-conductive steel would only be inserted after the hot zone exceeded 1400°F. I can also wrap a small pick-up coil around the bar's cooler end to monitor any fluctuation in the MMF created by the DC electromagnet. As the inner core's permittivity will change rapidly whenever the two exciter fields are applied, this coil should show those changes in a way that should convince most people. In theory, the power coming from the pickup coil should be AC with twice the frequency of the exciter input.

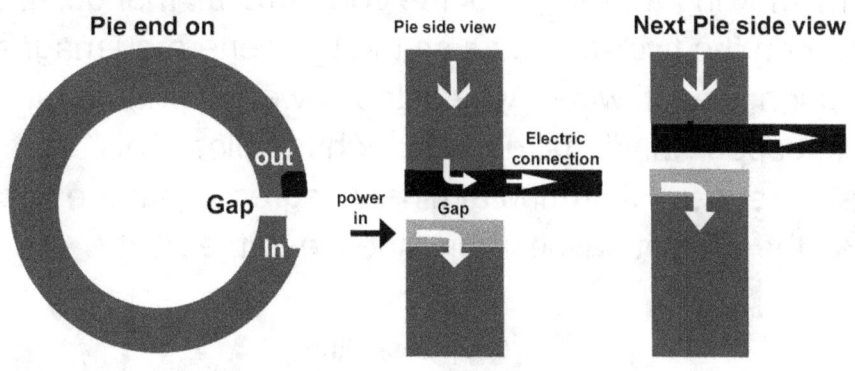

Above is a somewhat exaggerated view of the individual pies making up the inner core. Here, I'm trying to show how the current flows from one pie to the next. The white arrowheads indicate the direction of the electron current. Assume each pie is electrically insulated from the next except for the, welded in place, electric connection bar, forcing the current to circle each pie in the same direction. If I make these pies by cutting up a one inch diameter tube with a one-half inch hole in its center, it's going to require some fancy marching to form them correctly. I'm still debating this, and wondering if I should invest in the equipment necessary to drill a larger than one-inch hole, so I can use larger diameter pies, which would be easier to produce.

Once the pies are connected to one another, they will be wrapped in a small layer of electrical insulation *(the high temperature kind)* and then a single layer of heavy-gauge copper wire wrapped around them. This is to create the changing MMF, which should be in phase with the AC current flowing through the pies.

The whole core assembly should be five or six inches long to occupy most of the hot zone. It won't be a very strong assembly, but most of its rigidly should come from the copper windings and the insulation cones on each end should keep it in place.

Next, I'm showing a close-up of the core, *(the bit that gets hot)* which is placed between the two steel bars so the DC generated magnetic flux passes through it lengthwise. AC electric power, in phase with the AC in the surrounding copper windings, goes though the pies. This arrangement makes the current spiral around at ninety degrees from the copper windings AC MMF as the current passes from one pie to the next.

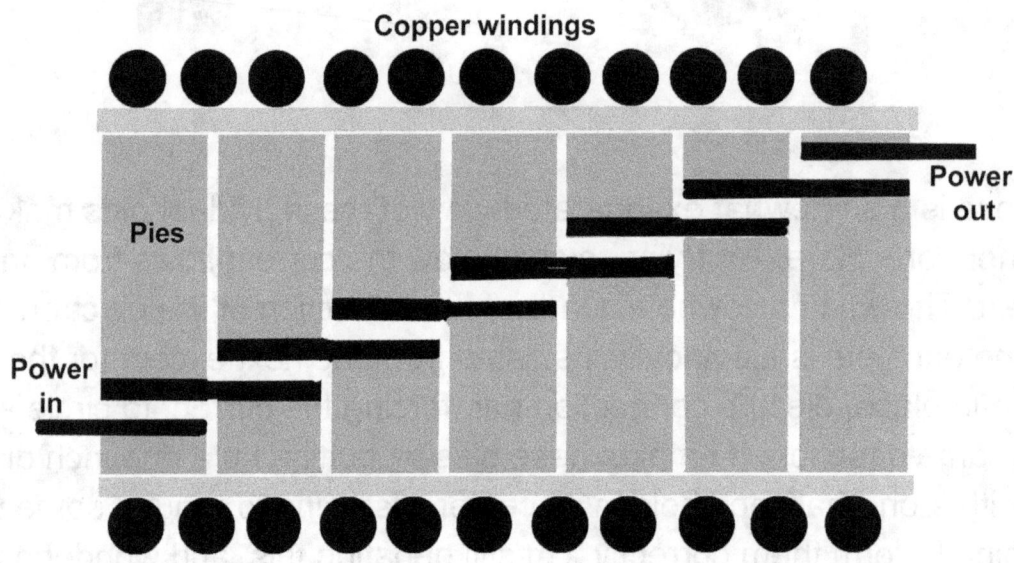

<u>A second idea</u>, which now sounds more reasonable, is to use two larger half-circle heaters.

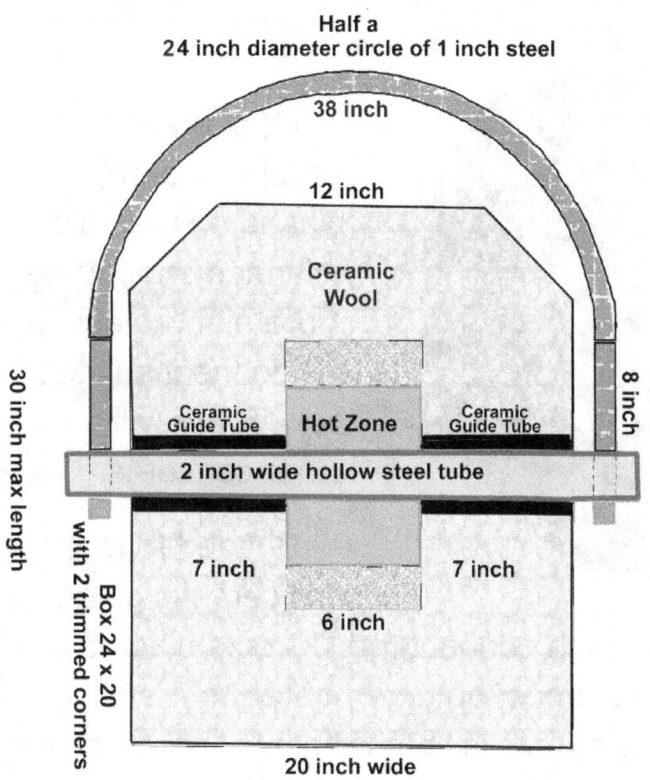

Above is a top-down view.

These are also available in 6 inches wide; however, their diameter is 12 inches. Each half circle gives 900 watts for a total of 1800. *(and overclocked, they might give me over 2000 watts for a short time.)*

I made a drawing of how the overall box would need to change and found it didn't make a significant difference.

Note the sides of the circular heaters are 2 inches think and the circular hot zone is 8 inches in diameter.

This will make it possible to use a 2-inch diameter tube for the core. I think a wider core will make it easier to fabricate the pies, especially when welding of one to the next. I'll need to make some clever jigs to machine each of them to one half-inch wide. Pretty much everything else will be about the same as I planned for the smaller square heaters.

Here's a side view that shows the proportions.

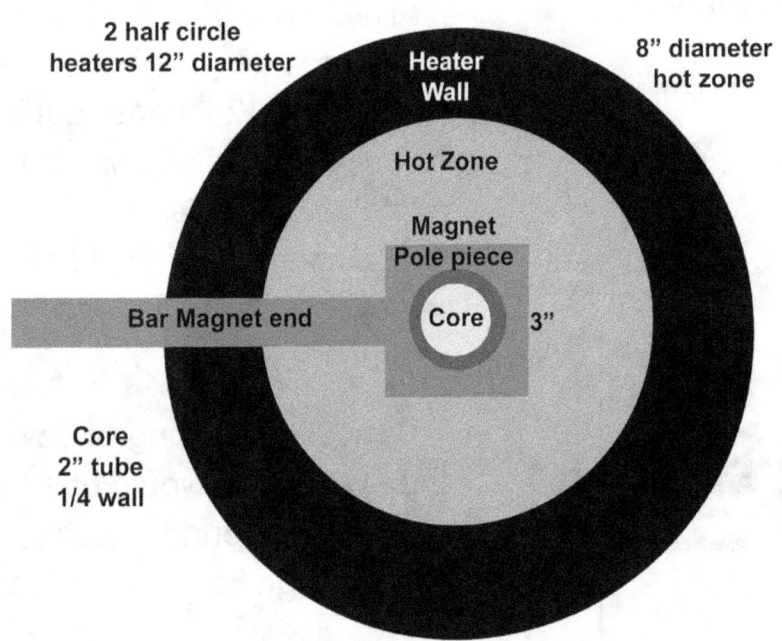

Of course, the circular heaters, placed on their side, will need supporting more elaborately inside the insulated box. There will be other concerns.

I'll need to make two ceramic guide tubes so the steel connector tubes, which connect the DC pole piece to the core, can be inserted after the core is almost up to temperature. Leaving them in place during the long heat up phase will both drain too much energy and might also cause the pickup coil to overheat.

This is my plan. I'm not sure how far along I'll get. Health and dwindling money supplies are always a concern. Part of the reason for this book is to provide a "jumping off" place for other tinkerers like myself. Or even, if I'm lucky, inspire some real research by someone with the resources to find out if electrical stress can change the Curie Point of a ferromagnetic material enough to create an efficient generator.

So, this is where I'm at today. Still planning to start from scratch with yet another prototype. I've reiterated enough times why I believe offbeat ideas like this should be examined in detail. All of that still holds true. If humanity intends to move forward, then we need to accept carbon-based energy sources can only be an intermittent step in our progress.

Today, humans use 18 trillion watts of energy every year and 80% comes from fossil fuel. *(That's 18 with 12 zeros.)* Fortunately, the sun provides many times that much power and will continue to do so for many more millions of years. We just have to make the commitment now, while there is still time to adjust, to find better ways to harness the sun's bounty.

Never forget, one advantage direct conversion of sunlight into electric power has over fossil fuel is that it isn't tied to the Carnot Cycle. Every heat source that begins with burning carbon must, by the laws of physics, waste around two thirds of the energy stored in the fuel. It's this gain that would allow ideas like this to compensate for the low overall power levels of sunlight. I still maintain an Electro-Hydrogen economy could lead us into a stable energy and climate future.

That's my hope. And I believe there are still enough people out there who care enough about our future generations to keep chasing this elusive dream.

Even if all you get is a whole book full of failures.

Thanks for reading my book, John.

About the Author:

I hale from England, born during the post World War Two years. Just before my thirteenth birthday, an Aunt and Uncle offered me the chance to immigrate to America. A few years later, on America's first Law Day, I became an American citizen. Becoming an American, remains one of the proudest accomplishments of my life.

After marrying the perfect women for me, a beautiful Florida cutie, we moved to Chicago to see snow and live in a big city. For a few fondly-remembered years our lives were happy and productive.

An accident left my wife in what doctors call a vegetative state and drove us back to Florida and family support. Being house-bound, I began tinkering, curious to see if an idea I'd had about an alternative energy source had any merit. Plus, as I'd always enjoyed writing stores, the hours of sitting at a bedside gave me the time to indulge in this lifelong desire.

I have published Four novels that you can buy online.

My Novels

Life commitments prevented me from pursuing writing seriously, however, after my wife's accident, I found myself with lots of spare time as I took care of her. When I couldn't go out to the shop and work on my generator idea, I sat at her bedside and continued my passion for writing. I've never made a serious push to become an established author, but my novels have won several awards and such feedback I've received has been positive and accompanied by praise.

The first series is a romantic fantasy. It's about a modern-day High School teacher forcibly relocated into a Bronze Age society. I call the trilogy "*Beneath the Jeweled Moon*." All my stories are written to stand alone. The only reason to read them in order is to avoid spoilers from the previous books.

I'm showing these covers in black and white to keep the cost of this book down. But they look much better in color.

Daughter of a Fallen Angel

Book one of *"Beneath the Jeweled Moon"*

A fallen angel decides her enslaved daughter, Sachi, needs a husband who can appreciate her good qualities and overlook her tiny-little flaw. Traveling to Earth, the angel discovers a perfect specimen on a sinking trawler and hopes she can get away with a little karmic interference.

Matthew Sharpe, an American descendent of the rebellious Jamaican Maroons, has major issues with waking up in a slave-owning culture on another world. However, he finds the cute slave girl, who's assigned to nurse him back to heath, the most desirable female he's ever encountered.

But the city is at war with a monster army, and one of the ruling witches knows of Sachi's true heritage. Fearful the girl might awaken to her power, she convinces her sisters that Matthew is an enemy half-breed sent to penetrate their defenses. Between avoiding death by the suspicious witches, fighting the attacking monsters, and dealing with a lover who's had enough of being a slave, Matthew learns some love affairs are more complicated than others.

Outcast City

Book two of *"Beneath the Jeweled Moon"*

They couldn't bring themselves to kill the evil witch. Now they must pay the price.

Matthew, Sachi and the ex-slaves have followed Taro's boatbuilder family into the snake-infested badlands with the dream of building a city where everyone is free. But winter approaches. As the colonists gather their meager harvest, the witch, Oki Nami, casts a spell to recapture the male slaves and leave their wives and children to die. The remaining colonist's only hope is Sachi and her undeveloped fallen-angel powers. Sachi can offer only one small chance, and it involves traveling to a post-apocalyptic earth with only five days to find a way to return before the women and children start dying.

Ex-high-school teacher Matthew Sharpe has mostly come to believe this strange magical land is real. Besides, building a city from scratch doesn't allow time for considering the impossibilities of magic. But his doubt returns after he ends up in a nightmare version of the Earth he left behind. Especially when he and Taro, once again face off against an army bent on their destruction. Unfortunately, this time their opponents have modern weapons.

Brother of a Fire Witch

Book three of *"Beneath the Jeweled Moon"*

Prince Haruko only wants to be left alone to build steam engines and mechanical devices, but his sister, the witch Tanoshi, has a problem. After turning people into frogs and single-handedly defeating an army, everyone fears and avoids her. But, as she approaches marriageable age, Haruko realizes she needs a husband, or face death from the Moon Goddess' curse.

Their lives become complicated after Haruko is ordered to the capitol to marry a noble's daughter who thinks anything mechanical is stupid. There is one answer, but can kids raised in a steampunk world survive in modern-day America? And can Tanoshi live without the power she'd known all her life?

Clericals, Courtesans and Superconductors.

Book one of *"The Rise and Fall of Synfood."*

I also wrote a series about our near future whose underlying theme concerns science and technology providing answers to our immediate physical needs while ignoring the more fundamental problems of human nature. I know that sounds a bit dystopian, but the story is mostly a romantic adventure with lots of action, suspense and flying machines in pieces on the ground. I call this series *"The Rise and Fall of Synfood."*

So far, I have only the first book in the Science-fiction series available.

Note, all my books are written to stand alone.

There are no cliff-hanger endings.

The story:

In 2137, a gene-altered supersmart knows she can out-think anyone in science and math. It's everything else that's a problem.

A Capitol bureaucrat, seeking revenge on a superior, transfers her to the office of the Secretary of Agriculture, a position for which she's not qualified.

Ruth then discovers she has two weeks to impress a cabinet member or end up living on the street. Her goal of providing a decent life for her extended family is further threatened when her new boss hints of his romantic interest.

But could a rich, aristocratic man understand a girl from the ass-end of a two-tier society? And would her heart break if she moved away from the couple who'd made her human? But she needs to keep in mind it takes only two clicks of a mouse to fire a clerical during their trial period. Even one who knows the secret of preventing an all-out war with China, the world's leading superpower.

These books are available on Amazon and all the better on-line book sellers.

Plus, you can always check out my website: **www.ferrogenerator.com**

Thanks, John

www.ingramcontent.com/pod-product-compliance
Lightning Source LLC
Chambersburg PA
CBHW081113180526
45170CB00008B/2824